Current Topics in Microbiology 133 and Immunology

Editors

A. Clarke, Parkville/Victoria · R.W. Compans,
Birmingham/Alabama · M. Cooper, Birmingham/Alabama
H. Eisen, Paris · W. Goebel, Würzburg · H. Koprowski,
Philadelphia · F. Melchers, Basel · M. Oldstone,
La Jolla/California · P.K. Vogt, Los Angeles
H. Wagner, Ulm · I. Wilson, La Jolla/California

This collection of studies was conceived as part of a two volume review of the subject. The contents of Volume 134 are listed below.

Arenaviruses

Genes, Proteins, and Expression

Edited by M. B. A. Oldstone

With 39 Figures

Springer-Verlag
Berlin Heidelberg New York
London Paris Tokyo

MICHAEL B. A. OLDSTONE, M.D.
Department of Immunology
Scripps Clinic and Research Foundation
10666 North Torrey Pines Road
La Jolla, CA 92037, USA

ISBN-13:978-3-642-71685-0   e-ISBN-13:978-3-642-71683-6
DOI: 10.1007/978-3-642-71683-6

2123/3130-543210

Table of Contents

Indexed in Current Contents

List of Contributors

You will find the addresses at the beginning of the respective contribution

The Arenaviruses – An Introduction

M.B.A. OLDSTONE

Viruses are generally studied either because they cause significant human, animal or plant disease or for their utility as materials to probe a basic phenomenon in biology, chemistry, genetics or molecular biology. Arenaviruses are unusually interesting in that they occupy both of these categories.

Arenaviruses cause severe human diseases known primarily as the hemorrhagic fevers occurring in South and Latin America (Bolivia: Machupo virus and Argentina: Junin virus) and in Africa (Lassa virus). Because such viruses produce profound disability and may kill the persons they infect, they are a source of economic hardship in the countries where they are prevalent. Further, they provide new problems for health care personnel owing to the narrowing of the world as visitors from many countries increasingly travel to and from these endemic areas. In addition, lymphocytic choriomeningitis virus (LCMV) can infect humans worldwide, although the illness is most often less disabling than those elicited by other arenaviruses. Yet LCMV is likely of greater concern to non-arena-virologists and experimentalists using tissue culture or animals, i.e., workers in molecular biology, cancer research, virology, immunobiology, etc., because normal appearing cultured cells or tissues and animals used for research may be persistently infected with LCMV without manifesting clinical disease or cytopathology and transmit that infection to laboratory workers (reviewed OLDSTONE and PETERS 1978). For example, HINMAN et al. (1975) recorded 48 cases among personnel in the radiotherapy department and vivarium at the University of Rochester School of Medicine. These persons had contact with Syrian hamsters into which tumors, unknown to be occultly infected with LCMV, were transferred. These tumors were obtained from an outside research supplier who distributed tumor cell lines to numerous laboratories primarily interested in SV-40 and polyoma virus research. In subsequent investigation of the 22 tumor lines in question, 13 yielded infectious LCMV. Infection was spread from the experimental area, presumably via air ducts. Similarly, the Communicable Disease Center, USA, has recorded multiple cases of LCMV infections in families scattered from the northeast coast of the United States (New York) to the western region (Reno, Nevada) originating from hamsters sold by a single supplier in the southeast (Florida). Several investigators found that their hybridomas making monoclonal antibodies were infected with LCMV. The mechanism was likely by the use of infected splenic feeder cells obtained

Department of Immunology, Scripps Clinic and Research Foundation, 10666 North Torrey Pines Road, La Jolla, CA 92307, USA

Current Topics in Microbiology and Immunology, Vol. 133

Table 1. Recognized Arenavirus diseases of man

Virus	Disease	Locality	Rodent reservoir and vector	Person to person transmission	Laboratory model of human infection
Lymphocytic choriomeningitis	Grippe, aseptic meningitis, occasional more severe forms of meningoencephalomyelitis	Probably originated in Europe, now worldwide	Mus musculus natural host; colonized rodents, particularly mice and hamster; dog?	Never documented	Mouse
Junin	Argentinian hemorrhagic fever	Circumscribed area of Argentina; Buenos Aires to northeast	Calomys musculinus, possibly others	Occasional	Guinea pig
Machupo	Bolivian hemorrhagic fever	Beni region of Bolivia	Calomys callosus	Occasional, particularly spouses; recognized hospital outbreak	Rhesus monkey
Lassa	Lassa fever	Western Africa	Mastomys nataliensis	Frequent; blood contamination	Squirrel monkey, Guinea pig strain 13 LCMV WE in guinea pig

from clinically healthy but persistently infected mice purchased from a commercial source. Hence, researchers studying such diverse areas as the molecular biology of SV40, biologic effects of chlamydia, immune responses of mice making hybridoma cells or production of ascites fluids have found their preparations contaminated with LCMV (LEHMANN-GRUBE 1973; WHO Bulletin 1975; GRIMWOOD 1983, 1985; VAN DER ZEIJST et al. 1983a, b). Thus, experimentalists may be exposed to the unexpected and potentially dangerous effects of LCMV owing to its non-cytolytic persistent infection in tissue culture cell lines and experimental or pet animals. The recognized arenavirus diseases of man, the vectors and laboratory models used for their study are listed in Table 1.

Study of the biology of LCMV has led to several major advances in contemporary virology and immunology. For example, research on both the acute and persistent infection of mice with LCMV led to the first descriptions of virus induced immunopathologic disease (ROWE 1954), of T cell mediated killing (COLE et al. 1973; MARKER and VOLKERT 1973) and of the two unique and separate signals necessary for recognition and lysis of virus infected targets by cytotoxic T lymphocytes (CTLs) (ZINKERNAGEL and DOHERTY 1974). Hence, study of the LCMV indicated that CTLs kill virus infected targets only when two conditions are met: recognition of virus specific determinants and matching of the major histocompatibility complex on the CTL and the infected target cell. This phenomenon of MHC restriction, one of the major tenets of contemporary immunology, was not only first described by experimentalists working with LCMV but also its implications were clearly recognized in this setting (reviewed by ZINKERNAGEL and DOHERTY 1979). Additionally, the components, kinetics, genetic control and deposition of immune complexes from the circulation into a variety of tissues including renal glomeruli, brain choroid plexus and blood vessel arteries were explored by utilizing the LCMV model (OLDSTONE and DIXON 1969). The conclusions were extended to work on numerous RNA and DNA virus infections as well as infections with other microbial agents (reviewed OLDSTONE 1975). Recently, by examination of its supposedly benign effect, LCMV was shown to cause disease not by means of the well-known anatomical viral induced destruction of infected cells, but rather by disordering the synthesis of such cells' differentiated products (OLDSTONE et al. 1982; reviewed OLDSTONE 1984). These studies in animals indicate that viruses may cause disease by altering the cells' physiology and function without causing their destruction. If these findings are applicable to human viruses, in vivo, we can anticipate much-needed understanding of the mechanisms by which dysfunctions of endocrine, nervous and immune systems occur.

Arenaviruses obtained their name from arenosus – Latin for sandy – on the basis of characteristic fine granularities seen inside the virion on ultrathin section (ROWE et al. 1970). To avoid confusion between areno and adeno viruses, the International Committee on Viral Nomenclature changed the name to arenaviruses (ROWE et al. 1970).

Recently, the ability of such viruses to cause persistent infection in cultured cells and animals, as well as their non-cytolytic nature, has been coupled with the wealth of information known about their biologic activity to inspire intensive investigation into the molecular control of viral genes and the molecular basis of arenavirus induced disease. It is the melding of newly obtained information on the viral chromosomes and gene structure, the proteins encoded by various viral genes and the ambisense (positive-genomic sense and negative-complementary sense) organization of the arenavirus genome – particularly as relates to the molecular basis of viral persistence, tissue tropism, viral gene regulation and mechanism of pathogenesis and biology of the virus – that led to the selection of arenaviruses for these two volumes in the Current Topics in Microbiology and Immunology series.

References

Cole GA, Prendergast RA, Henney CS (1973) In vitro correlates of LCM virus-induced immune response. In: Lehmann-Grube F (ed) Lymphocytic Choriomeningitis Virus and Other Arenaviruses. Springer, Berlin Heidelberg New York

Grimwood BG (1985) Viral contamination of a subline of *Toxoplasma gondii* RH. Infect Immun 50:917–918

Grimwood BG, O'Connor G, Gaafar HA (1983) Toxofactor associated with *Toxoplasma gondii* infection is toxic and teratogenic to mice. Infect Immun 42:1126–1135

Hinman AR, Fraser DW, Douglas RG Jr, Bowen GS, Kraus AL, Winkler WG, Rhodes WW (1975) Outbreak of lymphocytic choriomeningitis virus infections in medical center personnel. Am J Epidemiol 101:103–110

Lehmann-Grube F (ed) (1973) Lymphocytic Choriomeningitis Virus and Other Arenaviruses. Springer, Berlin Heidelberg New York

Marker O, Volkert M (1973) In vitro measurement of the time course of cellular immunity to LCM virus in mice. In: Lehmann-Grube F (ed) Lymphocytic Choriomeningitis Virus and Other Arenaviruses. Springer, Berlin Heidelberg New York

Oldstone MBA (1975) Virus neutralization and virus-induced immune complex disease: Virus-antibody union resulting in immunoprotection or immunologic injury – two different sides of the same coin. In: Progress in Medical Virology. Karger, Basel

Oldstone MBA (1984) Virus can alter cell function without causing cell pathology: Disordered function leads to imbalance of homeostasis and disease. In: Notking AL, Oldstone MBA (eds) Concepts in Viral Pathogenesis. Springer, New York

Oldstone MBA, Dixon FJ (1969) Pathogenesis of chronic disease associated with persistent lymphocytic choriomeningitis viral infection. I. Relationship of antibody production to disease in neonatally infected mice. J Exp Med 129:483–505

Oldstone MBA, Peters CJ (1978) Arenavirus infections of the nervous system. In: Handbook of Clinical Neurology. North-Holland, Amsterdam

Oldstone MBA, Sinha YN, Blount P, Tishon A, Rodriguez M, von Wedel R, Lampert PW (1982) Virus-induced alterations in homeostasis: Alterations in differentiated functions of infected cells in vivo. Science 218:1125–1127

Rowe WP (1954) Research Report NM 005 048.14.01. Nav Med Res Inst, Bethesda, Maryland

Rowe WP, Murphy FA, Bergold GH, Casals J, Hotchin J, Johnson KM, Lehmann-Grube F, Mims CA, Traub E, Webb PA (1970) Notes Arenaviruses: Proposed name for a newly defined virus group. J Virol 5:651–652

Van der Zeijst BAM, Bleumink N, Crawford LV, Swyryd EA, Stark GR (1983a) Viral proteins and RNAs in BHK cells persistently infected by lymphocytic choriomeningitis virus. J Virol 48:262–270

Van der Zeijst BAM, Noyes BE, Mirault M-E, Parker B, Osterhaus ADME, Swyryd EA, Bleumink N, Horzinek MC, Stark GR (1983b) Persistent infection of some standard cell lines by lymphocytic choriomeningitis virus: Transmission of infection by an intracellular agent. J Virol 48:249–261

World Health Organization (1975) International Symposium on Arenavirus Infections of Public Health Importance. Bull WHO 52:Nos. 4, 5, 6

Zinkernagel RM, Doherty PC (1974) Immunological surveillance against altered self-components by sensitized T lymphocytes in lymphocytic choriomeningitis. Nature (London) 251:547–548

Zinkernagel RM, Doherty PC (1979) MHC-restricted cytotoxic T cells: Studies on the biological role of polymorphic major transplantation antigens determining T-cell restriction specificity, function, and responsiveness. Adv Immunol 27:51–177

Arenavirus Gene Structure and Organization

D.H.L. Bishop[1] and D.D. Auperin[2]

1   Introduction

For rhabdoviruses, paramyxoviruses, orthomyxoviruses, and some members of the Bunyaviridae, all the proteins are translated from the viral-complementary RNA sequence. This is in contrast to the genetic strategy of picornaviruses, caliciviruses, coronaviruses, togaviruses, flaviviruses, and retroviruses, whose proteins are all translated from the viral RNA sequence (i.e., viral RNA, or sequences corresponding to viral RNA sequences, functions as an mRNA species). The former group of single-stranded RNA viruses are described as negative-stranded RNA viruses, the latter as positive-stranded RNA viruses. The negative-stranded RNA viruses have a virion RNA polymerase that is responsible for the synthesis of the viral-complementary mRNA species. Although the purified viral RNA of most of the positive-stranded viruses is infectious per se, this is not the case for retroviruses since they have a virion reverse transcriptase that forms an obligatory DNA intermediate during the initial stages of the viral infection process.

Arenaviruses have a genome consisting of two species of single-stranded RNA (Vezza et al. 1978), designated L (large) and S (small). The RNA is not infectious per se (Carter et al. 1974) and it does not appear to template the synthesis of protein when supplied to competent in vitro translation reactions (Leung et al. 1977). Viral RNA polymerase activities have been identified in Pichinde arenavirus preparations (Carter et al. 1974; Leung et al. 1979). Thus these data indicate that arenaviruses also have a negative-stranded coding strategy. Recent nucleotide sequence analyses of the S genome species of three members of the arenavirus family have shown that the 3′ half of the RNA sequence codes for a protein (the nucleoprotein, NP) in the viral-complementary

---

[1] NERC Institute of Virology, Mansfield Road, Oxford OX1 3SR, United Kingdom
[2] Centers for Disease Control, Atlanta, GA 30333, USA

Current Topics in Microbiology and Immunology, Vol. 133
© Springer-Verlag Berlin·Heidelberg 1987

sequence, the 5′ half codes for a protein (the glycoprotein precurser, GPC) in a viral-sense sequence. This strategy is described as an ambisense coding arrangement and is one of the subjects of this chapter.

2   Nucleotide Sequence and Coding Arrangement of the Arenavirus S RNA Species

The complete nucleotide sequences of the S RNA species of Pichinde arenavirus, lymphocytic choriomeningitis virus (LCMV) (WE strain) and Lassa fever virus, have recently been determined from partial RNA sequence data (AUPERIN et al. 1982a, b) and from analyses of representative clones of cDNA (AUPERIN et al. 1984a, b, 1986; CLEGG and ORAM 1985; ROMANOWSKI and BISHOP 1985; RO-MANOWSKI et al. 1985). The results indicate that all three viruses have a similar organization for their S RNA species. The viral and viral-complementary sequences of the S RNA of Pichinde arenavirus are shown in Fig. 1 together with the sequences of the encoded gene products (AUPERIN et al. 1984a, b). The S RNA is reported to be 3419 nucleotides long. The base composition of the viral RNA is 22.2% G, 22.6% C, 26.8% A, and 28.4% U, representing a size of approximately 1.1×10^6 daltons. Similar results have been obtained for LCM and Lassa fever viruses (CLEGG and ORAM 1985; ROMANOWSKI and BISHOP 1985; ROMANOWSKI et al. 1985; AUPERIN et al. 1986).

Genetic studies have shown that the arenavirus S RNA species codes for two proteins, NP and GPC (VEZZA et al. 1980; HARNISH et al. 1983; RIVIERE et al. 1985). Two methionine-initiated open reading frames that are sufficiently large to encode the two proteins have been identified among the six possible reading frames of the viral and viral-complementary sequences of the S RNA of Pichinde virus (AUPERIN et al. 1984a, b). One of the reading frames is in the viral-complementary RNA sequence and corresponds to the 3′ half of the viral RNA (Fig. 1a). The open reading frame begins at viral-complementary residue 84 and terminates at position 1766. It encodes a protein of 561 amino acids. The protein has been identified as the viral NP based on the observation that antibodies raised to one of its predicted peptides (near the amino terminus of the gene product) precipitate Pichinde NP (LEUNG et al. 1984). The calculated size (62984 daltons), composition, net positive charge (+9), and hydropathic character of the protein (e.g., the absence of hydrophobic domains at either the amino or carboxy termini of the protein) are also consistent with this conclusion (AUPERIN et al. 1984a).

The second S-coded open reading frame is in the 5′ half of the viral RNA sequence (Fig. 1b; AUPERIN et al. 1984b). It begins at viral RNA residue 52 and terminates at residue 1560. The gene product consists of 503 amino acids (size: 57.3×10^3 daltons) and contains two hydrophobic domains, one close to the amino terminus, the other close to the carboxy terminus (Fig. 2). These two regions of hydrophobic amino acids are similar to those of other viral glycoproteins (ROSE et al. 1980; SKEHEL and WATERFIELD 1975). In addition, the protein is rich in cysteine residues and potential asparagine-linked glycosylation sites (Fig. 2). Preliminary tryptic peptide sequence data with [^{35}S]cyteine-labelled tryptic peptides (M. GALINSKY and D.H.L. BISHOP, unpublished information) indicate that the Pichinde virus GP1 glycoprotein is derived from the amino terminus of GPC and that the GP2 comes from the carboxy terminus. Antibodies made to synthetic peptides from the amino terminal and carboxy terminal regions of the LCMV GPC gene product have recently been shown to react with LCMV GP1 and GP2 glycoproteins, respectively (M. BUCHMEIER, personal communication).

Shown in Fig. 3 are the flanking nucleotides that surround the NP, GPC, and L RNA translation initiation codons of the Pichinde, LCM, and Lassa fever arenaviruses (AUPERIN et al. 1985a, b, 1986; CLEGG and ORAM 1985; ROMANOWSKI and BISHOP 1985; ROMANOWSKI et al. 1985) compared to those determined by KOZAK (1978, 1984) to represent the consensus flanking sequence for translation initiation of proteins for eukaryotes (i.e., CCA/GCC*AUG*G). The data compiled for the indicated translation initiation codons of arenaviruses indicate that there are purines in the -1 and -3 positions (counting the *AUG* as $+1$, $+2$, and $+3$) and most of the $+4$ positions, with a strong preference for an A residue in the -3 position (Fig. 3). These data agree with Kozak's observation that an A or G residue at -3 is frequently found flanking the functional AUG initiation codon. It has been noted (P. YOUNG, personal communication) that in the leader sequence to the Pichinde, LCM, and Lassa fever viruses S and (where it is known) L mRNA species, there is a conserved six nucleotide sequence (GAUCCU, residues 10–15, see Fig. 1) that is complementary to a 3' terminal sequence (CUAGGA) present in 18 *S* ribosomal RNA (ATMADJA et al. 1984). This complementarity may be important in arenavirus mRNA-ribosome interactions and subsequent translation of the mRNA.

The S RNA nucleotide sequence data predict that the mRNA for the arenavirus NP gene must consist of a viral-complementary sequence corresponding to at least the 3' half of the viral S RNA. Similarly, the mRNA for the GPC gene must correspond to at least the 5' half of the viral S RNA. Northern blot analyses, using Pichinde or LCM virus-infected cell RNA preparations that were resolved into size classes by gel electrophoresis and hybridized to the appropriate nick-translated viral DNA or terminally labelled, single-stranded, DNA species representing the nucleoprotein genes, have identified subgenomic, viral-complementary, NP mRNA species corresponding to the 3' halves of the respective S RNA (AUPERIN et al. 1984b; ROMANOWSKI et al. 1985). The subgenomic mRNA has been translated in vitro into NP as identified by the size of the product and by immunoprecipitation analyses. Similar studies using strand-specific probes derived from the glycoprotein gene have detected in infected cell extracts a viral-sense, subgenomic RNA (putative GPC mRNA) corresponding to the 5' half of the genome (AUPERIN et al. 1984b). The subgenomic mRNA species present in Pichinde virus-infected cells apparently lack poly A tails on their 3' termini as evidenced by their inability to bind to oligo dT cellulose columns (AUPERIN et al. 1984b).

In summary, it has been found that the arenavirus NP is encoded in a viral-complementary, subgenomic mRNA species and that the GPC is encoded in a viral-sense, subgenomic mRNA species. The antiparallel arrangement of

```
                                                        M  S  D  N  I  P  S  F  R  W  V  Q  S
CGCACAGTGGATCCTAGGCGACACTAGATCACGCTGTACGTTCACTTCTTTACTGACTCAGAGGAAGTGTGAACAACTCCAAAATGTCCGACAATATCCCATCGTTCCGCTGGGTGCAAT
         10        20        30        40        50        60        70        80        90       100       110       120

     L  R  R  G  L  S  N  W  T  H  P  V  K  A  D  V  L  S  D  T  R  A  L  L  S  A  L  D  F  H  K  V  A  Q  V  G  R  M  V  R
CCCTTAGGAGGGGCTTGTCCAACTGGACCCATCCTGTGAAGGCTGATGTGCTGTCAGACACAAGAGCACTGTTGTCTGCTCTTGACTTTCACAAAGTTGCTCAAGTCAAAGAATGGTGC
        130       140       150       160       170       180       190       200       210       220       230       240

     K  D  K  R  T  D  S  D  L  T  K  L  R  D  M  N  K  E  V  D  A  L  M  N  M  R  S  V  G  R  D  N  V  L  K  V  G  G  L  A
GCAAAGATAAGAGGACTGATTCTGATCTGACCAAGCTGAGGGACATGAACAAAGAGGTTGATGCTCTGATGAATATGAGATCAGTCCAAAGAGATAATGTACTTAAAGTGGGAGGCTTAG
        250       260       270       280       290       300       310       320       330       340       350       360

     K  E  E  L  M  E  L  A  S  D  L  D  K  L  R  K  K  V  T  R  T  E  G  L  S  Q  P  G  V  Y  E  G  N  L  T  N  T  G  L  E
CTAAAGAGGAGCTCATGGAGCTTGCATCTGATTTGGACAAGTTAAGGAAGAAAGTCACCAGAACCGAAGGCCTGTCTCAGCCTGGTGTCTATGAAGGCAATCTTACAAACACTCAGCTGG
        370       380       390       400       410       420       430       440       450       460       470       480

     Q  R  A  E  I  L  R  S  M  G  F  A  N  A  R  P  A  G  N  R  D  G  V  V  K  V  W  D  I  K  D  N  T  L  L  I  N  G  F  G
AGCAAAGGGCAGAGATTCTCCGCTCGATGGGGTTTGCTAATGCCAGACCCGCAGGTAACAGAGATGGTGTTGTGAAGGTCTGGGACATCAAGGATAACACACTATTGATTAATCAATTTG
        490       500       510       520       530       540       550       560       570       580       590       600

     S  M  P  A  L  T  I  A  C  M  T  E  Q  G  G  E  Q  L  N  D  V  V  Q  A  L  S  A  L  G  L  L  Y  T  V  K  F  P  N  M  T
GATCAATGCCAGCCTTGACCATCGCCTGTATGACAGAGCAAGGGGGTGAGCAGCTTAATGATGTTGTCCAAGCACTGAGTGCACTTGGTTTGCTTTACACTGTCAAGTTCCCGAACATGA
        610       620       630       640       650       660       670       680       690       700       710       720

     D  L  E  K  L  T  G  G  H  S  A  L  K  I  I  S  H  E  P  S  A  L  N  I  S  G  Y  N  L  S  L  S  A  A  V  K  A  A  A  C
CAGATCTGGAAAAGCTCACACAACAACACAGCGCCCTAAAAATCATCAGCCATGAGCCATCAGCCTTGAATATCTCTGGGTATAACCTTAGTCTGTCTGCAGCAGTCAAAGCAGCTGCTT
        730       740       750       760       770       780       790       800       810       820       830       840

     M  I  D  G  G  N  M  L  E  T  I  G  V  K  P  S  M  F  S  T  L  I  K  S  L  L  G  I  K  N  R  E  G  M  F  V  S  T  T  P
GTATGATTGATGGTGGCAACATGCTTGAGACCATCCAAGTGAAGCCCTCTATGTTTAGCACCCTCATAAAAAGTCTACTGCAGATTAAGAACCGTGAGGGTATGTTTGTGAGTACTACAC
        850       860       870       880       890       900       910       920       930       940       950       960

     G  G  R  N  P  Y  E  N  L  L  Y  K  I  C  L  S  G  D  G  W  P  Y  I  G  S  R  S  G  V  G  G  R  A  W  D  N  T  T  V  D
CCGGGCAAAGGAACCCTTATGAAAAACCTGCTGTATAAAATCTGTCTCTCAGGGGATGGCTGGCCCTACATTGGCTCAAGATCCCAAGTTCAAGGGAGGGCTTGGGACAACACCACTGTGG
        970       980       990      1000      1010      1020      1030      1040      1050      1060      1070      1080

     L  D  S  K  P  S  A  I  G  P  P  V  R  N  G  G  S  P  D  L  K  G  I  P  K  E  K  E  D  T  V  V  S  S  I  G  M  L  D  P
ATTTGGACTCAAAGCCAAGTGCTATCCAGCCACCAGTAAGAAACGGTGGATCACCAGATCTCAAACAAATCCCTAAGGAAAAAGAAGACACTGTTGTGTCCTCAATTCAGATGCTTGATC
       1090      1100      1110      1120      1130      1140      1150      1160      1170      1180      1190      1200

     R  A  T  T  W  I  D  I  E  G  T  P  N  D  P  V  E  M  A  I  Y  G  P  D  T  G  N  Y  I  H  C  Y  R  F  P  H  D  E  K  S
CAAGAGCCACTACGTGGATTGATATTGAGGGGACACCAAATGACCCGGTGGAAATGGCCATCTACCAACCTGATACAGGCAACTACATACACTGTTATAGGTTTCCCCATGATGAAAAAT
       1210      1220      1230      1240      1250      1260      1270      1280      1290      1300      1310      1320

     F  K  E  G  S  K  Y  S  H  G  L  L  L  K  D  L  A  D  A  G  P  G  L  I  S  S  I  I  R  H  L  P  G  N  M  V  F  T  A  G
CCTTCAAAGAGCAGAGCAAGTATTCACATGGTCTCCTTTTAAAGGACTTGGCTGATGCTCAACCGGGTCTGATCTCCTCAATCATCAGACATTTGCCCCAGAACATGGTTTTCACTGCTC
       1330      1340      1350      1360      1370      1380      1390      1400      1410      1420      1430      1440

     G  S  D  D  I  I  R  L  F  E  M  H  G  R  R  D  L  K  V  L  D  V  K  L  S  A  E  G  A  R  T  F  E  D  E  I  W  E  R  Y
AAGGTTCAGATGATATAATCAGACTGTTCGAAATGCATGGAAGAAGAGATCTAAAAGTACTTGACGTGAAGCTCAGTGCAGAGCAGGCACGCACCTTTGAGGATGAGATCTGGGAGAGAT
       1450      1460      1470      1480      1490      1500      1510      1520      1530      1540      1550      1560
```

```
      N  G  L  C  T  K  H  K  G  L  V  I  K  K  K  K  K  G  A  V  G  T  T  A  N  P  H  C  A  L  L  D  T  I  M  F  D  A  T  V
ACAACCAACTCTGCACCAAGCACAAGGGCTTAGTCATAAAGAAGAAGAAGAAAGGAGCTGTACAAACCACTGCAAACCCCACTGTGCATTGCTTGACACCATCATGTTTGATGCAACAG
   1570      1580      1590      1600      1610      1620      1630      1640      1650      1660      1670      1680

      T  G  W  V  R  D  G  K  P  M  R  C  L  P  I  D  T  L  Y  R  N  N  T  D  L  I  N  L
TGACAGGCTGGGTCAGAGATCAGAAACCCATGAGATGTTTACCTATTGACACACTGTACAGGAACAACACAGACTTGATCAACCTCTGAGCTTAATCCTCGGAGGCCTCGACGTCACTCC
   1690      1700      1710      1720      1730      1740      1750      1760      1770      1780      1790      1800

CCTATTGGGGAGTGCCGTCGAGGCCCATGTCGGAAGCGGAGCTTATTTTCCCAACCTTACCCATTTGTAAGGTTTCTTTGGTATTTATAATACCCGCAGCTACACAGGGAGTTTCTTGTA
   1810      1820      1830      1840      1850      1860      1870      1880      1890      1900      1910      1920

ATCCTGTGTGGTTTCGGGCAACCATCACCAATGATGTGCCTATGAGTAGGTATTCCAACTATGTGGAGAAATACTGTAATGGTGTAAAAAACCAAAGACCAGAAGCATATATCTGTCAGT
   1930      1940      1950      1960      1970      1980      1990      2000      2010      2020      2030      2040

GCCAAAGGGGTTTTTCCTTGTCTTTCCTCATACTCTTTCATCAGCATCTCATTGTATAGATTTTGGCTCTCCCACAACCAATCATTCTTGAAATGCGTTTCATTGAGGTACGAGCCATTG
   2050      2060      2070      2080      2090      2100      2110      2120      2130      2140      2150      2160

TGAACCAGCCAACACTGCGGTAAAGAATGCCTTCCTGTGATAGTGTCATTGATGTACCAAAATTTTGTGTAGTTGCAGTAAGGAATTTTAGCAAGTTGTTTGAGACTGTTTCTAATCACA
   2170      2180      2190      2200      2210      2220      2230      2240      2250      2260      2270      2280

AGTGAGTCAGAAATGAGTCCGTTGATGGTCTTTTTGAAAAGATTCAGTGAATTCTCAACATTGAGTTGCAAGGTTTTGATGGCATTTTGATTGAAATCAAATAACCTCATTGTGTCGCAA
   2290      2300      2310      2320      2330      2340      2350      2360      2370      2380      2390      2400

AATTCCTCGTTGTGATCTTTGTTGCATTTTGCCATTACAGTGTTGTCGAAACATTTTATTCCAGCCCAGATAATGGCCCACTGCTCTAGACAGTAGCCACCTGGGACATGTTGCCCAGAG
   2410      2420      2430      2440      2450      2460      2470      2480      2490      2500      2510      2520

GAATCACTCAAGTCCCAAGTGAAGAAGCCAAGAAGTTTCCTGCTAACAGAACTATAAGCAGTTCTTTGAAGAGCCATTCTTATTGTTGCCATTGGTGTGTATGTGCAATGATTTTCCCAT
   2530      2540      2550      2560      2570      2580      2590      2600      2610      2620      2630      2640

GTGGTGTTCTGTATGATTAAGAAATTGTAATGTGTTCCACCCTCACAGTTTGTCAGTCTGCAAGTATCTCCACTGCAGTTGTTGAAGCACTTCCCAACCCATGCGATTTTAGGATCCCCG
   2650      2660      2670      2680      2690      2700      2710      2720      2730      2740      2750      2760

ATGATTTGAGCGAGTGATGCAATAAGATGTCTGCCAACCTCACCTCCTCTGTCCCCAACTGTCAAGTTGTATTGGATTAACACCCCAGCACCTTCAACTGTCTTACATCTGGCACCTATA
   2770      2780      2790      2800      2810      2820      2830      2840      2850      2860      2870      2880

TGACGAGTCACATGGAGCACATTGAAGTGTAGTTCATTGAGTAACCATTTCAATGTATGTCCTGCTTCCCTTGTCCTGTCACAATTGCTAATATTGCCATATCCAAGGCTCCCGATGTTG
   2890      2900      2910      2920      2930      2940      2950      2960      2970      2980      2990      3000

GAGAAGTTCCCACTAGTTTCATTAGCAATAGATGTGTTTGTCAAGGTCAGTTCAATTCCCCATGTCGTGTTGGATGGTCCTTTATAGTAATGGTGTGTGTTGTTCTTGCTGCATGATTGA
   3010      3020      3030      3040      3050      3060      3070      3080      3090      3100      3110      3120

GGCAGATTGTCAAACATTCGTGTGAGATTGAATTCAACATGGGTGAGATTGTGCCTTCTATCAATCATCATGCTGTCACAACTTCTGCCAGACAAAATGAGGAAGGTGACGAGCTGGAAA
   3130      3140      3150      3160      3170      3180      3190      3200      3210      3220      3230      3240

AGGCCACATCTCATCAGATTGACAAATCCTTTGACAATGCATAGAACTGAGACAATGATCAGAGCAACATTGAAGACCTCTTGCAGGACTTCAGGAATGGACTGGATCAAAGTCACAATT
   3250      3260      3270      3280      3290      3300      3310      3320      3330      3340      3350      3360

TGTCCCATCTTTGATCAAGAAATATGCGCGTCCAAGGTATGCCTAGGATCCCCGGTGCG
   3370      3380      3390      3400      3410
```

Fig. 1a

```
                         M  G  G  I  V  T  L  I  G  S  I  P  E  V  L  G  E  V  F  N  V  A  L
CGCACCGGGGGATCCTAGGCATACCTTGGACGCGCATATTTCTTGATCAAAGATGGGACAAATTGTGACTTTGATCCAGTCCATTCCTGAAGTCCTGCAAGAGGTCTTCAATGTTGCTCTG
     10        20        30        40        50        60        70        80        90       100       110       120

     I  I  V  S  V  L  C  I  V  K  G  F  V  N  L  M  R  C  G  L  F  Q  L  V  T  F  L  I  L  S  G  R  S  C  D  S  M  M  I  D
ATCATTGTCTCAGTTCTATGCATTGTCAAAGGATTTGTCAATCTGATGAGATGTGGCCTTTTCCAGCTCGTCACCTTCCTCATTTTGTCTGGCAGAAGTTGTGACAGCATGATGATTGAT
     130       140       150       160       170       180       190       200       210       220       230       240

     R  R  H  N  L  T  H  V  E  F  N  L  T  R  M  F  D  N  L  P  G  S  C  S  K  N  N  T  H  H  Y  Y  K  G  P  S  N  T  T  W
AGAAGGCACAATCTCACCCATGTTGAATTCAATCTCACACGAATGTTTGACAATCTGCCTCAATCATGCAGCAAGAACAACACACACCATTACTATAAAGGACCATCCAACACGACATGG
     250       260       270       280       290       300       310       320       330       340       350       360

     G  I  E  L  T  L  T  N  T  S  I  A  N  E  T  S  G  N  F  S  N  I  G  S  L  G  Y  G  N  I  S  N  C  D  R  T  R  E  A  G
GGAATTGAACTGACCTTGACAAACACATCTATTGCTAATGAAACTAGTGGGAACTTCTCCAACATCGGGAGCCTTGGATATGGCAATATTAGCAATTGTGACAGGACAAGGGAAGCAGGA
     370       380       390       400       410       420       430       440       450       460       470       480

     H  T  L  K  W  L  L  N  E  L  H  F  N  V  L  H  V  T  R  H  I  G  A  R  C  K  T  V  E  G  A  G  V  L  I  G  Y  N  L  T
CATACATTGAAAATGGTTACTCAATGAACTACACTTCAATGTGCTCCATGTGACTCGTCATATAGGTGCCAGATGTAAGACAGTTGAAGGTGCTGGGGTGTTAATCCAATACAACTTGACA
     490       500       510       520       530       540       550       560       570       580       590       600

     V  G  D  R  G  G  E  V  G  R  H  L  I  A  S  L  A  Q  I  I  G  D  P  K  I  A  W  V  G  K  C  F  N  N  C  S  G  D  T  C
GTTGGGGACAGAGGAGGTGAGGTTGGCAGACATCTTATTGCATCACTCGCTCAAATCATCGGGGATCCTAAAATCGCATGGGTTGGGAAGTGCTTCAACAACTGCAGTGGAGATACTTGC
     610       620       630       640       650       660       670       680       690       700       710       720

     R  L  T  N  C  E  G  G  T  H  Y  N  F  L  I  I  G  N  T  T  W  E  N  H  C  T  Y  T  P  M  A  T  I  R  M  A  L  G  R  T
AGACTGACAAACTGTGAGGGTGGAACACATTACAATTTCTTAATCATACAGAACACCACATGGGAAAATCATTGCACATACACACCAATGGCAACAATAAGAATGGCTCTTCAAAGAACT
     730       740       750       760       770       780       790       800       810       820       830       840

     A  Y  S  S  V  S  R  K  L  L  G  F  F  T  W  D  L  S  D  S  S  G  G  H  V  P  G  G  Y  C  L  E  G  W  A  I  I  W  A  G
GCTTATAGTTCTGTTAGCAGGAAACTTCTTGGCTTCTTCACTTGGGACTTGAGTGATTCCTCTGGGCAACATGTCCCAGGTGGCTACTGTCTAGAGCAGTGGGCCATTATCTGGGCTGGA
     850       860       870       880       890       900       910       920       930       940       950       960

     I  K  C  F  D  N  T  V  M  A  K  C  N  K  D  H  N  E  E  F  C  D  T  M  R  L  F  D  F  N  G  N  A  I  K  T  L  G  L  N
ATAAAATGTTTCGACAACACTGTAATGGCAAAAATGCAACAAAGATCACAACGAGGAATTTTGCGACACAATGAGGTTATTTGATTTCAATCAAAATGCCATCAAAACCTTGCAACTCAAT
     970       980       990      1000      1010      1020      1030      1040      1050      1060      1070      1080

     V  E  N  S  L  N  L  F  K  K  T  I  N  G  L  I  S  D  S  L  V  I  R  N  S  L  K  Q  L  A  K  I  P  Y  C  N  Y  T  K  F
GTTGAGAATTCACTGAATCTTTTCAAAAAGACCATCAACGGACTCATTTCTGACTCACTTGTGATTAGAAACAGTCTCAAACAACTTGCTAAAATTCCTTACTGCAACTACACAAAATTT
    1090      1100      1110      1120      1130      1140      1150      1160      1170      1180      1190      1200

     W  Y  I  N  D  T  I  T  G  R  H  S  L  P  G  C  W  L  V  H  N  G  S  Y  L  N  E  T  H  F  K  N  D  W  L  W  E  S  Q  N
TGGTACATCAATGACACTATCACAGGAAGGCATTCTTTACCGCAGTGTTGGCTGGTTCACAATGGCTCGTACCTCAATGAAACGCATTTCAAGAATGATTGGTTGTGGGAGAGCCAAAT
    1210      1220      1230      1240      1250      1260      1270      1280      1290      1300      1310      1320

     L  Y  N  E  M  L  M  K  E  Y  E  E  R  Q  G  K  T  P  L  A  L  T  D  I  C  F  W  S  L  V  F  Y  T  I  T  V  F  L  H  I
CTATACAATGAGATGCTGATGAAAGAGTATGAGGAAAGACAAGGAAAAACCCCTTTGGCACTGACAGATATATGCTTCTGGTCTTTGGTTTTTTACACCATTACAGTATTTCTCCACATA
    1330      1340      1350      1360      1370      1380      1390      1400      1410      1420      1430      1440

     V  G  I  P  T  H  R  H  I  I  G  D  G  C  P  K  P  H  R  I  T  R  N  S  L  C  S  C  G  Y  Y  K  Y  Q  R  N  L  T  N  G
GTTGGAATACCTACTCATAGGCACATCATTGGTGATGGTTGCCCGAAACCACACAGGATTACAAGAAACTCCCTGTGTAGCTGCGGGTATTATAAATACCAAAGAAACCTTACAAATGGG
    1450      1460      1470      1480      1490      1500      1510      1520      1530      1540      1550      1560

TAAGGTTGGGAAAATAAGCTCCGCTTCCGACATGGGCCTCGACGGCACTCCCCAATAGGGGAGTGACGTCGAGGCCTCCGAGGATTAAGCTCAGAGGTTGATCAAGTCTGTGTTGTTCCT
    1570      1580      1590      1600      1610      1620      1630      1640      1650      1660      1670      1680

GTACAGTGTGTCAATAGGTAAACATCTCATGGGTTTCTGATCTCTGACCCAGCCTGTCACTGTTGCATCAAACATGATGGTGTCAAGCAATGCACAGTGGGGGTTTGCAGTGGTTTGTAC
    1690      1700      1710      1720      1730      1740      1750      1760      1770      1780      1790      1800

AGCTCCTTTCTTCTTCTTCTTTATGACTAAGCCCTTGTGCTTGGTGCAGAGTTGGTTGTATCTCTCCCAGATCTCATCCTCAAAGGTGCGTGCCTGCTCTGCACTGAGCTTCACGTCAAG
    1810      1820      1830      1840      1850      1860      1870      1880      1890      1900      1910      1920

TACTTTTAGATCTCTTCTTCCATGCATTTCGAACAGTCTGATTATATCATCTGAACCTTGAGCAGTGAAAACCATGTTCTGGGGCAAATGTCTGATGATTGAGGAGATCAGACCCGGTTG
    1930      1940      1950      1960      1970      1980      1990      2000      2010      2020      2030      2040
```

```
AGCATCAGCCAAGTCCTTTAAAAGGAGACCATGTGAATACTTGCTCTGCTCTTTGAAGGATTTTTCATCATGGGGAAACCTATAACAGTGTATGTAGTTGCCTGTATCAGGTTGGTAGAT
   2050      2060      2070      2080      2090      2100      2110      2120      2130      2140      2150      2160

GGCCATTTCCACCGGGTCATTTGGTGTCCCCTCAATATCAATCCACGTAGTGGCTCTTGGATCAAGCATCTGAATTGAGGACACAACAGTGTCTTCTTTTTCCTTAGGGATTTGTTTGAG
   2170      2180      2190      2200      2210      2220      2230      2240      2250      2260      2270      2280

ATCTGGTGATCCACCGTTTCTTACTGGTGGCTGGATAGCACTTGGCTTTGAGTCCAAATCCACAGTGGTGTTGTCCCAAGCCCTCCCTTGAACTTGGGATCTTGAGCCAATGTAGGGCCA
   2290      2300      2310      2320      2330      2340      2350      2360      2370      2380      2390      2400

GCCATCCCCTGAGAGACAGATTTTATACAGCAGGTTTTCATAAGGGTTCCTTTGCCCGGGTGTAGTACTCACAAACATACCCTCACGGTTCTTAATCTGCAGTAGACTTTTTATGAGGGT
   2410      2420      2430      2440      2450      2460      2470      2480      2490      2500      2510      2520

GCTAAACATAGAGGGCTTCACTTGGATGGTCTCAAGCATGTTGCCACCATCAATCATACAAGCAGCTGCTTTGACTGCTGCAGACAGACTAAGGTTATACCCAGAGATATTCAAGGCTGA
   2530      2540      2550      2560      2570      2580      2590      2600      2610      2620      2630      2640

TGGCTCATGGCTGATGATTTTTAGGGCGCTGTGTTGTTGTGTGAGCTTTTCCAGATCTGTCATGTTCGGGAACTTGACAGTGTAAAGCAAACCAAGTGCACTCAGTGCTTGGACAACATC
   2650      2660      2670      2680      2690      2700      2710      2720      2730      2740      2750      2760

ATTAAGCTGCTCACCCCCTTGCTCTGTCATACAGGCGATGGTCAAGGCTGGCATTGATCCAAATTGATTAATCAATAGTGTGTTATCCTTGATGTCCCAGACCTTCACAACACCATCTCT
   2770      2780      2790      2800      2810      2820      2830      2840      2850      2860      2870      2880

GTTACCTGCGGGTCTGGCATTAGCAAACCCCATCGAGCGGAGAATCTCTGCCCTTTGCTCCAGCTGAGTGTTTGTAAGATTGCCTTCATAGACACCAGGCTGAGACAGGCCTTCGGTTCT
   2890      2900      2910      2920      2930      2940      2950      2960      2970      2980      2990      3000

GGTGACTTTCTTCCTTAACTTGTCCAAATCAGATGCAAGCTCCATGAGCTCCTCTTTAGCTAAGCCTCCCACTTTAAGTACATTATCTCTTTGGACTGATCTCATATTCATCAGAGCATC
   3010      3020      3030      3040      3050      3060      3070      3080      3090      3100      3110      3120

AACCTCTTTGTTCATGTCCCTCAGCTTGGTCAGATCAGAATCAGTCCTCTTATCTTTGCGCACCATTCTTTGGACTTGAGCAACTTTGTGAAAGTCAAGAGCAGACAACAGTGCTCTTGT
   3130      3140      3150      3160      3170      3180      3190      3200      3210      3220      3230      3240

GTCTGACAGCACATCAGCCTTCACAGGATGGGTCCAGTTGGACAAGCCCCTCCTAAGGGATTGCACCCAGCGGAACGATGGGATATTGTCGGACATTTTGGAGTTGTTCACACTTCCTCT
   3250      3260      3270      3280      3290      3300      3310      3320      3330      3340      3350      3360

GAGTCAGTAAAGAAGTGAACGTACAGCGTGATCTAGTGTCGCCTAGGATCCACTGTGCG
   3370      3380      3390      3400      3410  I
```

Fig. 1b

Fig. 1a, b. Nucleotide sequences of the viral-complementary (**a**) and viral (**b**) S RNA species of Pichinde arenavirus presented as the DNA sequence equivalents. The orientation of the sequences are from 5′ to 3′. The amino acid sequence of the gene product (NP) encoded in vc RNA sequence (**a**, residues 84–1766) is given above the corresponding nucleotide codons. The amino acid sequence of the gene product (GPC) encoded in RNA sequence (**b**, residues 52–1560) is given above the corresponding nucleotide codons

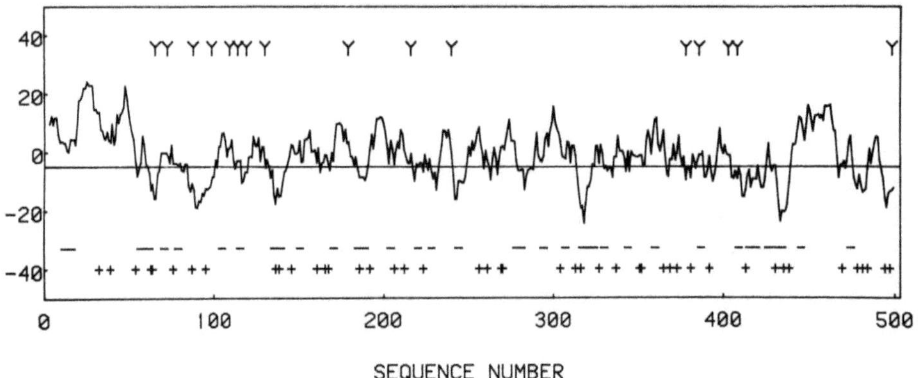

Fig. 2. Hydropathic plot, potential glycosylation sites, and charged amino acid distribution of the predicted GPC polypeptide of Pichinde arena-virus. The regions net hydrophobicity (areas *above* the center line), or net hydrophilicity (are *below* the center line) are displayed (KYTE and DOOLITTLE 1982) as well as the location of charged amino acids (R, K, D, E) and potential asparagine-linked glycosylation sites (*Y*, although whether all are employed for glycosylation is not known)

	Kozak consensus sequence	CC$^{A}_{G}$CCAUGG
Arenavirus S RNA	Viral-complementary RNA:	
	Pichinde NP gene	CCAAAAUGU
	LCM (WE) NP gene	ACAAAAUGU
	Lassa NP gene	CGACAAUGA
	Viral-sense RNA:	
	Pichinde GPC gene	CAAAGAUGG
	LCM (WE) GPC gene	AAAGGAUGG
	Lassa GPC gene	UUAAAAUGG
Arenavirus L RNA	Viral-complementary RNA:	
	LCM (WE) L gene	GCGCAAUGG

Fig. 3. Comparison of the nucleotides surrounding arenavirus translation initiation codons to those identified by KOZAK (1978, 1984) (CCA/GCCAUGG) as the most frequent flanking sequence for eukaryotic translation initiation codons

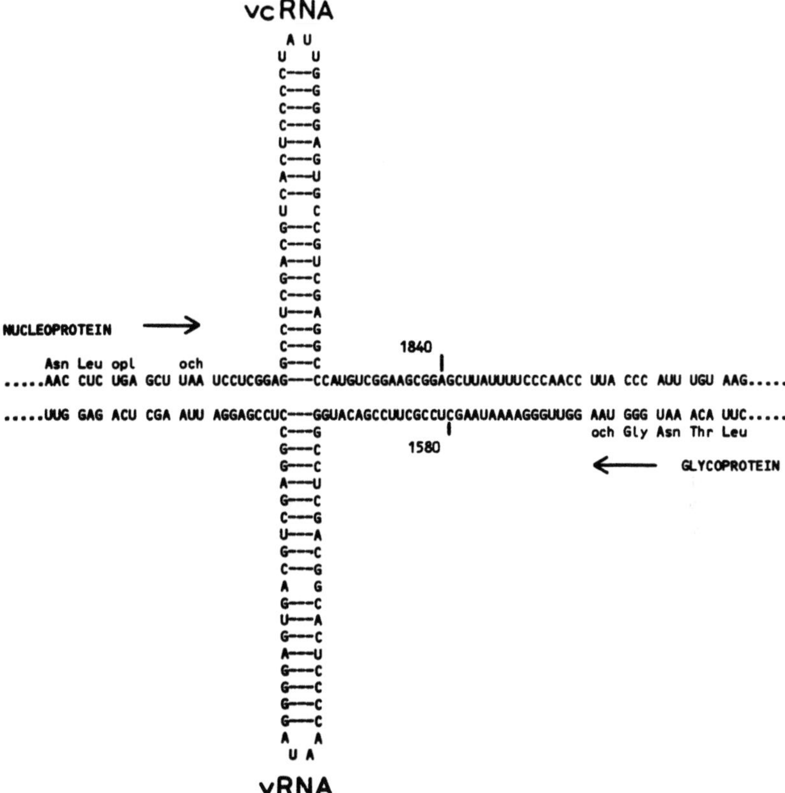

Fig. 4. The intergenic region of Pichinde virus S RNA species. The top sequence shows the viral-complementary (*vcRNA*) sequence and the carboxy terminal end of the coding region for the N. The bottom sequence shows the viral (*vRNA*) sequence and the carboxy terminal end of the coding region of the GPC gene product

the two genes on the S RNA is a novel strategy for RNA viruses. The term ambisense RNA has been used to describe it.

2.1 The Intergenic Region of the Arenavirus S RNA

The reading frames that code for the NP and GPC genes, as indicated in Fig. 1, do not overlap. For the S RNA of Pichinde virus, there is a short, noncoding, intergenic region of 87 nucleotides between the termination codons of the two genes. The nucleotide sequence within this intergenic region is such that a hairpin structure can be formed that would be stabilized by 14 G-C and 4 A-U base pairs (bp). The structure of the Pichinde virus intergenic region is illustrated in Fig. 4. Similar intergenic regions have been identified for LCM and Lassa fever viruses (ROMANOWSKI et al. 1985; AUPERIN et al. 1986).

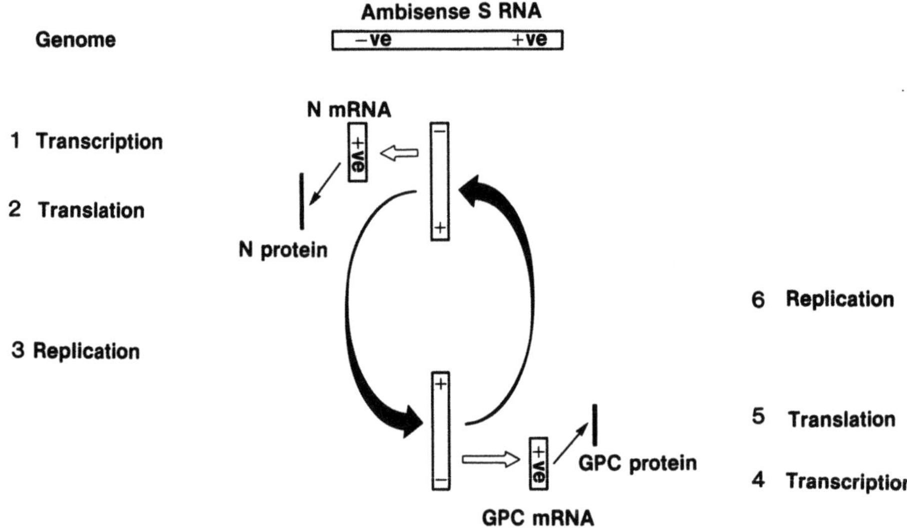

Fig. 5. The coding, transcription, and replication strategies of the arenavirus S RNA species

The unique aspect of the ambisense coding strategy is that it provides for the independent regulation of the synthesis of the NP and GPC genes. This is illustrated in Fig. 5. Presumably during a productive infection, transcription of the NP gene occurs immediately after virus adsorption and entry of the viral ribonucleoprotein into the host cell cytoplasm. However, the synthesis of the GPC mRNA cannot occur until viral RNA replication has commenced and full length, replicative intermediate, viral-complementary RNA is available as a template for transcription of the GPC gene. The independent regulation of the expression of these two genes may be important in the establishment and maintenance of persistently infected cells. GIMENEZ and COMPANS (1980) have shown that in cells persistently infected with Tacaribe virus the viral GPC is present only at very low levels, although the NP is abundant. Similar observations have been made for Pichinde virus-infected cells (V. ROMANOWSKI and D.H.L. BISHOP, unpublished data). These observations may be explained if, unlike the NP mRNA synthesis, GPC mRNA synthesis is specifically reduced or inhibited in such cells. One mechanism for selective inhibition could be that there are only small amounts of replicative intermediate, viral-complementary RNA species in those cells, or that such species are not used for GPC mRNA synthesis but are preferentially used for viral RNA replication.

Considering the size of the NP and GPC mRNA species, it is probable that the intergenic region functions as a transcription terminator for both genes. Evidence to support this notion has come from oligonucleotide annealing studies (M. GALINSKI and D.H.L. BISHOP, unpublished data). How during RNA replication the viral polymerase avoids this signal is not known.

```
                   1        10        20        30        40    46
Pichinde S vRNA   3'  GCGUGUCACCUAGGAUCCGCUGUGAUCUAGUGCGACAUGCAAGUGA
                      ******************               *  *  *   *  *

LCM S vRNA        3'  GCGUGUCACCUAGGAUCCGUAAACUAACGCGAAAAUAAACCUUUAA
                      ******************   ***   *    ***        *

Lassa S vRNA      3'  GCGUGUCACCUAGGAUCCGAUAACCUAACGCGAAACGUGAUUAGUU
                      ***** * ***********            ***      *  * *

Pichinde S vcRNA  3'  GCGUGGCCCCUAGGAUCCGUAUGGAACCUGCGCGUAUAAAGAACUA
                      ******************* *             *  **    **

LCM S vcRNA       3'  GCGUGGCCCCUAGGAUCCGAAAAACCUAACGCGAAAGGAAAUCCUG
                      ******************* **** * ******  * *     *  *

Lassa S vcRNA     3'  GCGUGGCCCCUAGGAUCCGUAAAAACCAACGCGUUAAGUUCACAGG
```

Fig. 6. The 3′ terminal nucleotide homologies of the viral and viral-complementary RNA sequences of Pichinde, LCM, and Lassa fever viruses aligned according to their termini (AUPERIN et al. 1984a, b, 1986; CLEGG and ORAM 1985; ROMANOWSKI and BISHOP 1985; ROMANOWSKI et al. 1985). The *asterisks* identify homologous nucleotides. Additional homologies exist if one incorporates gaps in the sequences, in particular for residues 21–35 of the LCMV and Lassa fever virus vRNA and for residues 20–33 of their vcRNA species

2.2 The 3′ and 5′ Terminal Nucleotide Sequences of Arenavirus L and S RNA Species

It has been shown that the 3′ and 5′ termini of the S genome RNA of Pichinde arenavirus are complementary for some 19 nucleotides with two mismatches (Fig. 1, residues 6 and 8; see AUPERIN et al. 1984b). Similar data have been obtained for the S RNA of LCMV (ROMANOWSKI et al. 1985). The complementarity between the terminal nucleotide sequences presumably reflects a consensus sequence required for the initiation of mRNA transcription and RNA replication at the 3′ termini of both the viral and viral-complementary (replicative intermediate) RNA species. It has been reported that there is an extra G nucleotide on the end of the cDNA clones of Pichinde virus corresponding to the 5′ terminus of the S RNA (AUPERIN et al. 1984b). This nucleotide was thought to be a cloning artifact, since its presence interfered with the exact alignment of the complementary termini. A similar observation has been made for Lassa fever virus (AUPERIN et al. 1986) but was not observed for LCMV (WE strain, ROMANOWSKI et al. 1985).

A comparison of the 3′ terminal nucleotide sequences of the viral and viral-complementary S RNA species of Pichinde, LCMV, and Lassa fever viruses shows that the 3′ terminal 19 nucleotides of the vRNA species are exactly conserved as are the 3′ terminal 19 nucleotides of the vcRNA species (Fig. 6). The three mismatches between the terminal nucleotide sequences of the respective viral and viral-complementary RNA species of each virus and their conservation among the three different arenaviruses suggests that the sequence differences of the 3′ and 5′ ends have a function that has been maintained through evolution. One possible explanation is that the 3′ terminal nucleotide sequence present on the viral and viral-complementary RNA templates may function in determining the efficiency with which mRNA transcription and/or viral RNA replication is initiated from either template. Such a mechanism could provide the basis for the mRNA regulation discussed above.

3 The Arenavirus L RNA

Unlike the S RNA, for the L genome RNA species of arenaviruses, there is only a limited amount of nucleotide sequence data available. ROMANOWSKI and BISHOP (1985) have published the nucleotide sequence for the 3′ terminal 1123 nucleotides of the L RNA of LCMV (WE strain). The data indicate that a viral gene product is encoded in the L viral-complementary RNA sequence beginning at nucleotide residue 33. The reading frame encoding this gene product remains open for the duration of the sequence determined, encoding a minimum, therefore, of 43×10^3 daltons of protein. Northern blot analyses using the cloned cDNA as a probe identified viral-size RNA species but failed to identify any L subgenomic RNA species. It is presumed that the viral RNA polymerase is encoded on the L RNA. Whether this reading frame encodes other gene products is not known. It is also not known whether the L RNA exhibits an ambisense coding strategy.

References

Atmadja J, Brimacombe R, Maden BEH (1984) *Xenopus laevis* 18S ribosomal RNA: experimental determination of secondary structural elements, and locations of methyl groups in the secondary structure model. Nucleic Acids Res 12:2649–2667

Auperin DD, Dimock K, Cash P, Rawls WE, Bishop DHL (1982a) Analyses of the genomes of prototype Pichinde arenavirus and a virulent derivative of Pichinde Munchique: evidence for sequence conservation at the 3′ termini of their viral RNA species. Virology 116:363–367

Auperin DD, Compans RW, Bishop DHL (1982b) Nucleotide sequence conservation at the 3′ termini of the virion RNA species of New World and Old World arenaviruses. Virology 121:200–203

Auperin DD, Galinsky M, Bishop DHL (1984a) The sequences of the N protein gene and intergenic region of the S RNA of Pichinde arenavirus. Virology 134:208–219

Auperin DD, Romanowski V, Galinsky M, Bishop DHL (1984b) Sequencing studies of Pichinde arenavirus S RNA indicate a novel coding strategy, an ambisense viral S RNA. J Virol 52:897–904

Auperin DD, Saso D, McCormick JBM (1986) Nucleotide sequence of the glycoprotein gene and intergenic region of Lassa fever virus. (manuscript in preparation)

Carter MF, Biswal N, Rawls WE (1974) Polymerase activity of Pichinde arenavirus. J Virol 13:577–583

Clegg JCS, Oram JD (1985) Molecular cloning of Lassa virus RNA: nucleotide sequence and expression of the nucleoprotein gene. Virology 144:363–372

Gimenez HB, Compans RW (1980) Defective interfering Tacaribe virus and persistent infected cells. Virology 107:229–239

Harnish DG, Dimock K, Bishop DHL, Rawls WE (1983) Gene mapping in Pichinde virus: assignment of viral polypeptides to genomic L and S RNAs. J Virol 46:638–641

Kozak M (1978) How do eucaryotic ribosomes select initiation regions in messenger RNAs. Cell 15:1109–1123

Kozak M (1984) Point mutations close to the AUG initiator codon affect the efficiency of translation of rat preproinsulin in vivo. Nature 308:241–246

Kyte J, Doolittle RF (1982) A simple method for displaying the hydropathic character of a protein. J Mol Biol 157:105–132

Leung W-C, Ghosh HP, Rawls WE (1977) Strandedness of Pichinde virus RNA. J Virol 22:235–237

Leung W-C, Leung MFKL, Rawls WE (1979) Distinctive RNA transcriptase, polyadenylic acid polymerase, and polyuridylic acid polymerase activities associated with Pichinde virus. J Virol 30:98–107

Leung W-C, Ramsingh A, Jing G, Mong K, Taneja AK, Hodges RS (1984) Subgenomic RNA of arenavirus Pichinde and its implication in regulation of viral gene expression in productive infection and persistence. Academic, New York

Riviere Y, Ahmed R, Southern PJ, Buchmeier MJ, Dutko FJ, Oldstone MBA (1985) The S RNA segment of lymphocytic choriomeningitis virus codes for the nucleoprotein and glycoproteins 1 and 2. J Virol 53:966–968

Romanowski V, Bishop DHL (1985) Conserved sequences and coding of two strains of lymphocytic choriomeningitis virus (WE and ARM) and Pichinde arenavirus. Virus Res 2:35–51

Romanowski V, Matsura Y, Bishop DHL (1985) Complete sequence of the S RNA of lymphocytic choriomeningitis virus (WE strain) compared to that of Pichinde arenavirus. Virus Res 3:101–114

Rose JK, Welch WJ, Sefton BM, Esch FS, Ling NC (1980) Vesicular stomatitis virus glycoprotein is anchored in the viral membrane by a hydrophobic domain near the COOH terminus. Proc Natl Acad Sci USA 77:3884–3888

Skehel JJ, Waterfield MD (1975) Studies of the primary structure of the influenza virus hemagglutinin. Proc Natl Acad Sci USA 72:93–97

Vezza AC, Clewley JP, Gard GP, Abraham NZ, Compans RW, Bishop DHL (1978) Virion RNA species of the arenaviruses Pichinde, Tacaribe, and Tamiami. J Virol 26:485–497

Vezza AC, Cash P, Jahrling P, Eddy G, Bishop DHL (1980) Arenavirus recombination: the formation of recombinants between prototype Pichinde and Pichinde Munchique viruses and evidence that arenavirus S RNA codes for N polypeptide. Virology 106:250–260

Sequence Comparison Among Arenaviruses

P.J. Southern[1] and D.H.L. Bishop[2]

1 Introduction

Contemporary techniques for DNA cloning and rapid nucleotide sequencing have made significant contributions to the present understanding of gene structure and regulation (see, for example, Gluzman and Shenk 1983; Gluzman 1985). Cloning has allowed the amplification of genetic material that may be difficult or impossible to propagate by conventional tissue culture methods (e.g., hepatitis B virus and several papillomaviruses) and provides unlimited material for hybridization probes and nucleotide sequencing. Information derived from sequencing has identified primary gene structures, amino acid sequences for protein coding regions, and regulatory signals that appear to be widespread in eukaryotic genomes (Gilbert 1981; Sanger 1981). Knowledge of a nucleotide sequence does not automatically provide answers to all outstanding questions, but the information does represent useful progress towards such answers.

The genomic sequences of different individual members from a virus family provide insight towards understanding common and distinctive features at the molecular level and may eventually reveal patterns of virus evolution (Reanney 1982; Holland 1984). Already, substantial information has been accumulated for the S segments of arenaviridae, and this review will consider three different viruses: Pichinde (Auperin et al. 1984), LCM (Romanowski and Bishop 1985; Romanowski et al. 1985; Southern et al. 1986b), and Lassa fever (Clegg and

[1] Department of Immunology, Scripps Clinic and Research Foundation, 10666 North Torrey Pines Road, La Jolla, CA 92037, USA
[2] NERC Institute of Virology, Mansfield Road, Oxford OX1 3SR, United Kingdom

Current Topics in Microbiology and Immunology, Vol. 133
© Springer-Verlag Berlin · Heidelberg 1987

ORAM 1985; AUPERIN et al. 1986) and will include comparisons between two different strains of both LCM and Lassa fever viruses.

2 Structural Organization of the Genomic S RNA Segment

The unusual "ambisense" character of the arenavirus genomic S RNA segment has been reviewed extensively by BISHOP and AUPERIN (this volume). All of the S sequence information currently available is consistent with the following general scheme: 3' nucleoprotein (NP) coding region and 5' glycoprotein (GPC) coding region. The NP mRNA is complementary to the genome, whereas the putative GPC mRNA is in the sense of the genome. The NP and GPC coding regions do not overlap and transcription termination for the mRNAs apparently occurs in a hairpin region located between the two genes (AUPERIN et al. 1984). The 3' and 5' termini of the S RNA segments contain 19 residues of complementary sequence that may represent the binding site for the viral polymerase. There is extensive sequence similarity between the 3' ends of the genomic S and L segments which supports the notion of a polymerase binding site in this region (AUPERIN et al. 1982a, b). It is anticipated that the 5' end of the L RNA will be complementary to the 3' L terminal sequence but this has not yet been confirmed by direct sequencing. Apart from sequence conservation of the termini, the genomic L and S RNA segments apparently do not contain extensive regions of homologous sequence. Weak cross-hybridization has been detected between L probes and S-derived RNAs in blotting experiments, but, as the regions involved spanned the carboxyl terminus of the putative L polymerase and the central portion of the GP2 coding sequence, it has not yet been possible to evaluate the significance of this observation. Information derived from cDNA clones covering approximately 50% of the genomic L RNA segment suggests that the L RNA represents a classical negative-stranded RNA segment at least in the region encoding the high molecular weight L protein (putative polymerase molecule). Further cloning and sequencing is required to determine whether additional proteins may be encoded within the L segment.

3 Experimental Approaches to Cloning and Sequencing

Once conditions had been established for efficient propagation of virus strains in vitro, then relatively large amounts of purified virion RNAs were available for analysis. The RNA components of the virion (principally viral genomic sense L and S RNAs and host 28 S, 18 S, and 5.8/5 S ribosomal RNA) (reviewed by PEDERSEN 1979) could be separated either by velocity gradient centrifugation or more effectively by electrophoresis in denaturing agarose gels. Oligonucleotide fingerprinting of purified genomic L and S RNA segments provided clear indications that these RNA species were comprised of largely divergent sequences and revealed differences among the commonly studied laboratory strains of LCMV (DUTKO and OLDSTONE 1983). Sequences from the 3′ termini of L and S RNAs were determined by 3′ end labeling of purified RNA segments in vitro followed by conventional chemical/enzymatic degradation techniques and separation on thin denaturing acrylamide gels. The resultant information revealed 3′ terminal similarities between L and S RNA segments of an individual virus and established sequence similarities at the termini of genomic RNA segments of different arenaviruses.

These direct RNA sequencing methods provided information for the synthesis of sequence-specific oligonucleotides which have served as primers for reverse transcriptase in cDNA cloning experiments.

4 Isolation and Characterization of cDNA Clones

Different approaches to reverse transcription and cloning have produced similar cDNA libraries derived from S genomic RNA segments. Techniques for priming reverse transcription have included sequence specific oligonucleotides, dT_{10-18} used in conjunction with 3′ polyadenylation of purified viral RNAs and random calf thymus primers. In general, it has been difficult to derive long (>2 kb) S clones and this perhaps reflects interference within the central hairpin region which may act as a major barrier to reverse transcriptase. Once inserted into a prokaryotic vector (most commonly the plasmid pBR322), the cloned arenavirus cDNA sequences appear to be maintained without rearrangement.

There is a major limitation with the cDNA cloning methods, in that any clone is derived from a single RNA molecule and therefore may not be entirely representative of the population. This is a serious concern when dealing with the genomes of RNA viruses that may undergo high mutation rates and therefore any given RNA preparation may contain several sequence variants (see, for example, SCHUBERT et al. 1984; DOMINGO et al. 1985). A more representative sequence of viral genomic RNA segments can be obtained from independently derived overlapping clones or from direct RNA sequencing using synthetic oligonucleotide primers distributed throughout the genome (GHOSH et al. 1980). Both of these approaches have been employed to confirm the accuracy and representative nature of the sequence derived from the S RNA segment of LCMV Armstrong (ARM) strain.

5 Analysis of Information Derived from Nucleotide Sequences

There is extensive conservation of genome organization among the arenavirus family members that have been examined to date. There are minor differences in the lengths of the genomic S RNA segment (for example Pichinde virus 3419 bases, LCMV WE 3375 bases) which are manifest in minor differences both within the coding regions and within nontranslated regions. It appears that approximately 6% of the genomic S segment constitutes nontranslated sequence. Precise mapping information for the 5′ and 3′ termini of the NP and GPC subgenomic mRNA species is not presently available. However, the AUG translation initiation codons in general lie 50–80 bases internal to the 3′ terminus of the appropriate template RNA, suggesting the possible presence of relatively short 5′ nontranslated regions. (Lassa NP mRNA falls slightly outside of this size estimate because the AUG codon is located at nucleotides 103–105.) Preliminary information from S1 nuclease mapping suggests that the LCMV ARM GPC mRNA initiates at or very close to the 3′ end of the full-length genomic complementary template RNA (SOUTHERN, unpublished results). The 3′ ends of mRNA species have been positioned in the proximity of the intergenic hairpin but identification of the precise termini awaits direct RNA sequencing. The interpretation of RNA hybridization experiments suggests that the mRNAs do not extend significantly beyond the hairpin into the other coding region but it is not clear whether the transcripts terminate (a) before the hairpin, (b) within the hairpin, or (c) immediately on the distal side of the hairpin. Perhaps inclusion of a hairpin structure at the 3′ ends of arenavirus mRNAs serves to stabilize the mRNAs and accounts for the apparent absence of the 3′ polyA tract which is present in the majority of eukaryotic messages. Clearly, future efforts will need to address the regulatory mechanism(s) that discriminate between transcription and replication.

The available nucleotide sequences have been used to predict amino acid sequences for the arenavirus NP and GPC. Homologies between different NP and GPC sequences have been established using a standard Align Program and a summary of the results is presented in Table 1. The actual alignments for the NP and GPC sequences are shown in the Appendix. There are minor differences in numbers of amino acid residues contained within the structural proteins NP and GPC, but blocks of conserved residues can readily be identified. Such conserved areas may indicate functional domains within the structural proteins, and regions of divergence must be involved with the unique characteristics of individual viruses. The conserved blocks appear to be distributed throughout the NP and GPC molecules and in some cases are flanked by amino acid residues which now have divergent character or which appear simply not to be present in one of the viral proteins. Such arrangements have implications for the patterns of virus evolution, suggesting either sequence duplication followed by rapid mutation or small, in-phase deletions; both events could arise as consequences of inaccurate replication. The ARM and WE strains of LCMV are clearly very closely related, yet these two viruses show remarkably different biological properties. Variations in their tropism and pathogenetic potential (RIVIERE et al. 1985a, b, c) apparently must reside in single or very limited amino acid substitutions. The ARM and WE strains also show differences with respect to neutralizing sites within GP1 (PAREKH and BUCHMEIER 1986; see also p. 41, this volume. The protein sequence comparisons suggest that Lassa is more closely related to LCMV then to Pichinde virus and that Pichinde virus shows the same distant relationship to LCMV as to Lassa fever virus. Conservation of sequence is significantly higher for GP2 than for GP1, as indicated by the higher number of identical amino acids and the reduced number of breaks introduced in the process of computer alignment. Similar findings had previously

Table 1. Amino acid homologies between arenavirus structural proteins[a]

Viral protein	Percentage homologous residues
Nucleoprotein	
PV/LCMV WE	50.5
PV/LCMV ARM	50.6
PV/Lassa	51.4
LCMV WE/LCMV ARM	95.7
LCMV WE/Lassa	63.3
LCMV ARM/Lassa	62.4
Glycoprotein	
PV/LCMV WE	44.6
PV/LCMV ARM	45.9
PV/Lassa	47.6
LCMV WE/LCMV ARM	93.6
LCMV WE/Lassa	61.3
LCMV ARM/Lassa	61.7

PV, Pichinde virus; *LCMV WE*, lymphocytic choriomeningitis virus WE strain; *LCMV ARM*, lymphocytic choriomeningitis virus Armstrong strain; *Lassa*, Lassa fever virus; *nucleoprotein*, GA 391 Nigerian strain; *glycoprotein*, Josiah strain.

[a] Amino acid sequences were compared using the Align program from the National Biomedical Research Foundation, Georgetown University Medical Center, Washington, D.C. Sequence alignments are shown in the Appendix.

been predicted from tryptic peptide mapping and cross-reactivity studies with an extensive panel of monoclonal antibodies (BUCHMEIER et al. 1980, 1981; BUCHMEIER 1984). Homology at the nucleotide level is in the range of 80%–85% for LCMV ARM and WE strains with frequent degeneracy occurring in the third base positions of codons. Transitions ($C \leftrightarrow T$ and $A \leftrightarrow G$) occur more frequently than transversions ($C \leftrightarrow A$ or G and $T \leftrightarrow A$ or G).

We have exploited the nucleotide sequence divergence of the LCMV ARM, Pasteur, and WE strains in the hybridization screening of reassortant (i.e., mixed genotype) viruses (see p. 5, this volume). Under standard conditions for hybridization, the ARM cDNA probes provided very effective discrimination between the homologous L and S ARM genomic RNAs and the heterologous Pasteur or WE L and S RNAs. The ARM cDNA probes do hybridize efficiently to the lymphotropic clone 13 variant that was derived by a single passage in vivo from the ARM stock (AHMED et al. 1984). Cloning and sequencing experiments are currently in progress to assess the extent of change between the parental ARM and the clone 13 variant (SALVATO and OLDSTONE, unpublished observations).

On a more specific level, the alignment of nucleotide sequences provides the opportunity to look for short stretches of homology which could be selected

```
   I N P N M S C D D V V F G I      LCMV Arm

   V N P N M S C D D V V F G I      LCMV WE

   C L A F S Y M D D V V L G A      Hepatitis B Virus

   L I L L Q Y V D D L L L A A      Murine Leukemia Virus

a  L K M I A Y G D D V I A S Y      Polio Virus


     G I N S L F S R F R            LCMV Arm

     G I N S F F G R F R            LCMV WE

b    S I N P L F P R F L            VSV
```

Fig. 1. a Homology around a conserved pair of aspartic acid (D-D) residues in the LCMV L protein and other viral polymerases. **b** Homology between LCMV L protein and VSV polymerase

as potential oligonucleotide probes to recognize multiple members of the arenavirus family. Equally, oligonucleotide probes can be selected which should be totally specific for an individual virus. In selecting sequences in the latter category, we have identified breaks in amino acid alignments and selected nucleotide sequences that "cross" the break, further ensuring specificity to the particular reagent.

Computer comparisons of the arenavirus nucleotide and amino acid sequences have also revealed some interesting similarities with other sequences in the databases. In the short stretch of overlapping amino acid sequence for the WE and ARM L protein, there is a conserved pair of aspartic acid residues in a hydrophobic pocket (Fig. 1a). This structure has previously been recognized in widely divergent viral polymerases and may be indicative of a functionally conserved site (KAMER and ARGOS 1984; SOUTHERN et al. 1986a). Also, we have noticed that from this same region there is a short stretch of amino acids with substantial homology to a region of the VSV polymerase (SCHUBERT et al. 1984; Fig. 1b). Again, the homology may be indicative of conserved function in the appropriate region of the two polymerase molecules.

6 Base Composition and Codon Usage

The available sequence information indicates that the Pichinde, LCMV, and Lassa viruses all have approximately 55% A/U in the genomic S segment. The arenavirus NP and GPC coding regions show a general, nonrandom codon usage where codons containing the dinucleotide CpG sequences are significantly underrepresented. This finding applies in general to sequences of eukaryotic chromosomal genes, and methylation of the C residue in CpG has been postulated as a general regulatory mechanism for eukaryotic gene expression (WIGLER et al. 1981). The possible significance of low CpG content in LCMV RNA sequences remains to be evaluated.

7 Concluding Remarks

Comparisons of arenavirus genomic S sequences indicate significant conservation between viruses, detected in widely separated regions. These findings establish foundations for speculation about the character of a common ancestral virus and will permit molecular analysis of the evolutionary process(es) occurring within this family of RNA viruses.

The potential now exists, through synthetic peptide technologies (LERNER 1982; p. 41 this volume), to generate arenavirus family-specific reagents. There were previous indications of this possibility from the peptide mapping and monoclonal antibody studies performed by BUCHMEIER and colleagues but the prediction of synthetic peptides from nucleotide sequences allows precise targeting of immunological reagents to defined regions of the arenavirus structural proteins.

In the future, sequence analysis of arenavirus L segments should begin to provide information on the properties of the viral polymerase/replicase and the regulatory mechanism to discriminate between replication and transcription. Already the L segment has been associated with LCMV pathogenesis in adult guinea pigs, and other situations may well be described in which the onset of disease can be linked to an L-derived protein, or that a particular replicase molecule supports more efficient and/or more extensive viral replication, thereby causing disease. As the L segment contains approximately two-thirds of the arenavirus genetic information, it is likely that more surprises are in store.

Acknowledgments. This is publication number 4362-IMM from the Department of Immunology, Scripps Clinic and Research Foundation, La Jolla, CA. This work was supported in part by USPHS grants NS-12428 and AG-04342 and by USAMRIID contract C-3013. The findings in this report are not to be construed as an official Departement of the Army position unless so designated by other authorized documents. We thank DAVID AUPERIN, CHRIS CLEGG, VICTOR ROMANOWSKI, and MANDAL SINGH whose cloning and sequencing work has contributed significantly to this review, and we acknowledge the excellent secretarial assistance of Ms. GAY L. SCHILLING.

Appendix

The following pages contain pairwise comparisons of the available amino acid sequences for arenavirus nucleoproteins and glycoproteins. This information was generated using the Align program from National Biomedical Research Foundation, Georgetown University Medical Center, Washington, D.C. Gaps (indicated by *dashes*) are introduced to maximize the alignment of amino acids and residues in common are indicated in the line below the individual amino acid sequences.

A1 Nucleoprotein ARM/WE

```
         1    5         10        15        20        25        30        35        40        45        50        55
ARM      M S L S K E V K S F Q W T Q A L R R E L Q S F T S D V K A A V I K D A T N L L N G L D F S E V S N V Q R I M R
WE       M S L S K E V K S F Q W T Q A L R R E L Q G F T S D V K A A V I K D A T S L L N G L D F S E V S N V Q R I M R
Common   M S L S K E V K S F Q W T Q A L R R E L Q   F T S D V K A A V I K D A T   L L N G L D F S E V S N V Q R I M R

         56   60        65        70        75        80        85        90        95        100       105       110
ARM      K E K R D D K D L Q R L R S L N Q T V H S L V D L K S T S K K N V L K V G R L S A E E L M S L A A D L E K L K
WE       K E R R D D K D L Q R L R S L N Q T V H S L V D P K S T S K K N V L K V G R L S A E E L M T L A A D L E K L K
Common   K E   R D D K D L Q R L R S L N Q T V H S L V D   K S T S K K N V L K V G R L S A E E L M   L A A D L E K L K

         111  115       120       125       130       135       140       145       150       155       160       165
ARM      A K I M R S E R P Q A S G V Y M G N L T T Q Q L D Q R S Q I L Q I V G M R K P Q Q G A S G V V R V W D V K D S
WE       A K I M R T E R P Q A S G V Y M G N L T A Q Q L D Q R S Q I L Q M V G M R R P Q Q G A S G V V R V W D V K D S
Common   A K I M R   E R P Q A S G V Y M G N L T   Q Q L D Q R S Q I L Q   V G M R   P Q Q G A S G V V R V W D V K D S

         166  170       175       180       185       190       195       200       205       210       215       220
ARM      S L L N N Q F G T M P S L T M A C M A K Q S Q T P L N D V V Q A L T D L G L L Y T V K Y P N L N D L E R L K D
WE       S L L N N Q F G T M P S L T M A C M A K Q S Q T P L N D V V Q A L T D L G L L Y T V K Y P N L S D L E R L K D
Common   S L L N N Q F G T M P S L T M A C M A K Q S Q T P L N D V V Q A L T D L G L L Y T V K Y P N L   D L E R L K D

         221  225       230       235       240       245       250       255       260       265       270       275
ARM      K H P V L G V I T E Q Q S S I N I S G Y N F S L G A A V K A G A A L L D G G N M L E S I L I K P S N S E D L L
WE       K H P V L G V I T E Q Q S S I N I S G Y N F S L G A A V K A G A A L L H G G N M L E S I L I K P S N S E D L L
Common   K H P V L G V I T E Q Q S S I N I S G Y N F S L G A A V K A G A A L L   G G N M L E S I L I K P S N S E D L L

         276  280       285       290       295       300       305       310       315       320       325       330
ARM      K A V L G A K R K L N M F V S D Q V G D R N P Y E N I L Y K V C L S G E G W P Y I A C R T S I V G R A W E N T
WE       K A V L G A K K K L N M F V S D Q V G D R N P Y E N I L Y K V C L S G E G W P Y I A C R T S V V G R A W E N T
Common   K A V L G A K   K L N M F V S D Q V G D R N P Y E N I L Y K V C L S G E G W P Y I A C R T S   V G R A W E N T

         331  335       340       345       350       355       360       365       370       375       380
ARM      T I D L T S E K P A V N S P R P A P G A A G P P Q V G L S Y S Q T M L L K D L M G G I D P N A P T W I D I E G
WE       T I D L T N E K L V A N S S R P V P G A A G P P Q V G L S Y S Q T M L L K D L M G G I D P N A P T W I D I E G
Common   T I D L T   E K       N S   R P   P G A A G P P Q V G L S Y S Q T M L L K D L M G G I D P N A P T W I D I E G

         386  390       395       400       405       410       415       420       425       430       435       440
ARM      R F N D P V E I A I F Q P Q N G Q F I H F Y R E P V D Q K Q F K Q D S K Y S H G M D L A D L F N A Q P G L T S
WE       R F N D P V E I A I F Q P Q N G Q F I H F Y R E P T D Q K Q F K Q D S K Y S H G M D L A D L F N A Q A G L T S
Common   R F N D P V E I A I F Q P Q N G Q F I H F Y R E P   D Q K Q F K Q D S K Y S H G M D L A D L F N A Q   G L T S

         441  445       450       455       460       465       470       475       480       485       490       495
ARM      S V I G A L P Q G M V L S C Q G S D D I R K L L D S Q N R K D I K L I D V E M T R E A S R E Y E D K V W D K Y
WE       S V I G A L P Q G M V L S C Q G S D D I R K L L D S Q N R R D I K L I D V E M T K E A S R E Y E D K V W D K Y
Common   S V I G A L P Q G M V L S C Q G S D D I R K L L D S Q N R   D I K L I D V E M T   E A S R E Y E D K V W D K Y

         496  500       505       510       515       520       525       530       535       540       545
ARM      G W L C K M H T G I V R D K K K K E I T P H C A L M D C I I F E S A S K A R L P D L K T V H N I L P H D L I F
WE       G W L C K M H T G V V R D K K K K E I T P H C A L M D C I I F E S A S K A R L P D L K T V H N I L P H D L I F
Common   G W L C K M H T G   V R D K K K K E I T P H C A L M D C I I F E S A S K A R L P D L K T V H N I L P H D L I F

         551  555       560
ARM      R G P N V V T L
WE       R G P N V V T L
Common   R G P N V V T L
```

A2 Nucleoprotein ARM/PV


```
       1   5    10   15   20   25   30   35   40   45   50   55
ARM    MSLSKEVKSFQWTQALRRELQSFTSDVKAAVIKDATNLLNGLDFSEVSNVQRIMR
PV     --MSDNIPSFRWVQSLTRGLSNWTHPVKADVLSDTRALLSALDFHKVAQVQRMVR
Common    S    SF W Q L R L   T  VKA V  D   LL  LDF  V  VQR  R

       56  60   65   70   75   80   85   90   95   100  105  110
ARM    KEKRDDKDLQRLRSLNQTVHSLVDLKSTSKKNVLKVGRLSAEELMSLAADLEKLK
PV     KDKRTDSDLTKLRDMNKEVDALMNMRSVQRDNVLKVGGLAKEELMELASDLDKLR
Common K KR D DL  LR  N  V  L    S    NVLKVG L  EELM LA DL KL

       111 115  120  125  130  135  140  145  150  155  160  165
ARM    AKIMRSERPQASGVYMGNLTTQQLDQRSQILQIVGMRKPQQGAS--GVVRVWDVK
PV     KKVTRTEGLSQPGVYEGNLTNTQLEQRAEILRSMGFANARPAGNRDGVVKVWDIK
Common  K  R E     GVY GNLT  QL QR  IL   G           GVV VWD K

       166 170  175  180  185  190  195  200  205  210  215  220
ARM    DSSLLNNQFGTMPSLTMACMAKQSQTPLNDVVQALTDLGLLYTVKYPNLNDLERL
PV     DNTLLINQFGSMPALTIACMTEQGGEQLNDVVQALSALGLLYTVKFPNMTDLEKL
Common D  LL NQFG MP LT ACM  Q    LNDVVQAL  LGLLYTVK PN  DLE L

       221 225  230  235  240  245  250  255  260  265  270  275
ARM    KDKHPVLGVITEQQSSINISGYNFSLGAAVKAGAALLDGGNMLESILIKPSNSED
PV     TQQHSALKIISHEPSALNISGYNLSLSAAVKAAACMIDGGNMLETIQVKPSMFST
Common    H  L  I    S  NISGYN SL AAVKA A   DGGNMLE I  KPS

       276 280  285  290  295  300  305  310  315  320  325  330
ARM    LLKAVLGAKRKLNMFVSDQVGDRNPYENILYKVCLSGEGWPYIACRTSIVGRAWE
PV     LIKSLLQIKNREGMFVSTTPGQRNPYENLLYKICLSGDGWPYIGSRSQVQGRAWD
Common L K  L  K    MFVS   G RNPYEN LYK CLSG GWPYI  R    GRAW

       331 335  340  345  350  355  360  365  370  375  380  385
ARM    NTTIDLTSEKPAVNSPRPAPGAAGPPQVGLSYSQTMLLKDLMGGIDPNAPTWIDI
PV     NTTVDLDSKPSAIQPPVRNGGSPDLKQIPKEKEDTVVSSIQM--LDPRATTWIDI
Common NTT DL S   A   P    G     Q       T      M   DP A TWIDI

       386 390  395  400  405  410  415  420  425  430  435  440
ARM    EGRFNDPVEIAIFQPQNGQFIHFYREPVDQKQFKQDSKYSHGMDLADLFNAQPGL
PV     EGTPNDPVEMAIYQPDTGNYIHCYRFPHDEKSFKEQSKYSHGLLLKDLADAQPGL
Common EG  NDPVE AI QP  G  IH YR P D K FK  SKYSHG L DL  AQPGL

       441 445  450  455  460  465  470  475  480  485  490  495
ARM    TSSVIGALPQGMVLSCQGSDDIRKLLDSQNRKDIKLIDVEMTREASREYEDKVWD
PV     ISSIIRHLPQNMVFTAQGSDDIIRLFEMHGRRDLKVLDVKLSAEQARTFEDEIWE
Common  SS I  LPQ MV   QGSDDI  L     R D K  DV    E  R  ED  W

       496 500  505  510  515  520  525  530  535  540  545  550
ARM    KYGWLCKMHTGIVRDKKKKEIT-----PHCALMDCIIFESASKARLPDLKTVHNI
PV     RYNQLCTKHKGLVIKKKKKGAVQTTANPHCALLDTIMFDATVTGWVRDQKPMR-C
Common  Y  LC H G V  KKKK        PHCAL D I F         D K

       551 555  560  565  570
ARM    LPHDLIFRG-PNVVTL
PV     LPIDTLYRNNTDLINL
Common LP D   R       L
```

A3 Nucleoprotein ARM/LV

```
          1         5        10        15        20        25        30        35        40        45        50        55
ARM       M S L S K E V K S F Q W T Q A L R R E L Q S F T S D V K A A V I K D A T N L L N G L D F S E V S N V Q R I M R
LV        M S A S K E V R S F L W T Q S L R R E L S G Y C S N I K L Q V V K D A Q A L L H G L D F S E V S N V Q R L M R
Common    M S   S K E V   S F   W T Q   L R R E L         S     K     V   K D A     L L   G L D F S E V S N V Q R   M R

          56        60        65        70        75        80        85        90        95       100       105       110
ARM       K E K R D D K D L Q R L R S L N Q T V H S L V D L K S T S K K N V L K V G R L S A E E L M S L A A D L E K L K
LV        K Q K R D D G D L K R L R D L N Q A V N N L V E L K S T Q Q K S V L R V G T L S S D D L L I L A A D L E K L K
Common    K   K R D D   D L   R L R   L N Q   V     L V   L K S T     K   V L   V G   L S       L     L A A D L E K L K

         111       115       120       125       130       135       140       145       150       155       160       165
ARM       A K I M R S E R P Q A S G V Y M G N L T T Q Q L D Q R S Q I L Q I V G M R K P Q Q G A S G - - - - V V R V W D
LV        S K V T R T E R P L S S G V Y M G N L S S Q Q L D Q R R A L L N M I G M T G V S G G G K G A S D G I V R V W D
Common      K     R   E R P     S G V Y M G N L     Q Q L D Q R       L       G M           G     G           V R V W D

         166       170       175       180       185       190       195       200       205       210       215       220
ARM       V K D S S L L N N Q F G T M P S L T M A C M A K Q S Q T P L N D V V Q A L T D L G L L Y T V K Y P N L N D L E
LV        V K N A E L L N N Q F G T M P S L T L A C L T K Q G Q V D L N D A V Q A L T D L G L I Y T A K Y P N S S D L D
Common    V K       L L N N Q F G T M P S L T   A C     K Q   Q     L N D   V Q A L T D L G L   Y T   K Y P N     D L

         221       225       230       235       240       245       250       255       260       265       270       275
ARM       R L K D K H P V L G V I T E Q Q S S I N I S G Y N F S L G A A V K A G A A L L D G G N M L E S I L I K P S N S
LV        R L S Q S H P I L N M I D T K K S S L N I S G Y N F S L G A A V K A G A C M L D G G N M L E T I K V S P Q T M
Common    R L       H P   L     I         S S   N I S G Y N F S L G A A V K A G A     L D G G N M L E   I       P

         276       280       285       290       295       300       305       310       315       320       325       330
ARM       E D L L K A V L G A K R K L N M F V S D Q V G D R N P Y E N I L Y K V C L S G E G W P Y I A C R T S I V G R A
LV        D G I L K S I L K V K K S L G M F V S D T P G E R N P Y E N I L Y K I C L S G D G W P Y I A S R T S I V G R A
Common          L K     L     K     L   M F V S D     G   R N P Y E N I L Y K   C L S G   G W P Y I A   R T S I V G R A

         331       335       340       345       350       355       360       365       370       375       380       385
ARM       W E N T T I D L T S E K P A V N S P R P A P G A A G P P Q - - - V G L S Y S Q T M L L K D L M G - G I D P N A
LV        W E N T V V D L E Q D N K P Q K I G N G G S N K S L Q S A G F A A G L T Y S Q L M T L K D F K C F N L I P N A
Common    W E N T     D L                                 G L   Y S Q   M   L K D               P N A

         386       390       395       400       405       410       415       420       425       430       435       440
ARM       P T W I D I E G R F N D P V E I A I F Q P Q N G Q F I H F Y R E P V D Q K Q F K Q D S K Y S H G M D L A D L F
LV        K T W M D I E G R P E D P V E I A L Y Q P S S G C Y V H F F R E P T D L K Q F K Q D A K Y S H G I D V T D L F
Common      T W   D I E G R     D P V E I A     Q P     G       H F   R E P   D   K Q F K Q D   K Y S H G   D     D L F

         441       445       450       455       460       465       470       475       480       485       490       495
ARM       N A Q P G L T S S V I G A L P Q G M V L S C Q G S D D I R K L L D S Q N R K D I K L I D V E M T R E A S R E Y
LV        A A Q P G L T S A V I E A L P R N M V I T C Q G S E D I R K L L E S Q G R R D I K L I D I T L S K A D S R K F
Common      A Q P G L T S   V I   A L P     M V     C Q G S   D I R K L L   S Q   R   D I K L I D             S R

         496       500       505       510       515       520       525       530       535       540       545
ARM       E D K V W D K Y G W L C K M H T G I V R D K K K K - - - - - E I T P H C A L M D C I I F E S A S K A R L P D L K
LV        E N A V W D Q F K D L C H M H T G V V V E K K K K R G G K E E I T P H C A L M D C I M F D A A V S G G L - D A K
Common    E     V W D         L C   M H T G   V     K K K K           E I T P H C A L M D C I   F     A         L   D   K

         551       555       560       565       570
ARM       T V H N I L P H D L I F R G - - P N V V T L
LV        V L R V V L P R D M V F R T S T P K V V L -
Common              L P   D     F R         P   V V
```

A4 Nucleoprotein WE/PV

```
        1       5         10        15        20        25        30        35        40        45        50        55
WE      M S L S K E V K S F Q W T Q A L R R E L Q G F T S D V K A A V I K D A T S L L N G L D F S E V S N V Q R I M R
PV      - - M S D N I P S F R W V Q S L T R G L S N W T H P V K A D V L S D T R A L L S A L D F H K V A Q V Q R M V R
Common        S       S F   W   Q   L   R   L           T       V K A   V       D           L L       L D F       V       V Q R       R

        56      60        65        70        75        80        85        90        95        100       105       110
WE      K E R R D D K D L Q R L R S L N Q T V H S L V D P K S T S K K N V L K V G R L S A E E L M T L A A D L E K L K
PV      K D K R T D S D L T K L R D M N K E V D A L M N M R S V Q R D N V L K V G G L A K E E L M E L A S D L D K L R
Common  K     R   D   D L     L R     N     V     L         S         N V L K V G   L     E E L M   L A   D L   K L

        111     115       120       125       130       135       140       145       150       155       160       165
WE      A K I M R T E R P Q A S G V Y M G N L T A Q Q L D Q R S Q I L Q M V G M R - - R P Q Q G A S G V V R V W D V K
PV      K K V T R T E G L S Q P G V Y E G N L T N T Q L E Q R A E I L R S M G F A N A R P A G N R D G V V K V W D I K
Common    K     R T E         G V Y   G N L T     Q L   Q R     I L       G         R P           G V V   V W D   K

        166     170       175       180       185       190       195       200       205       210       215       220
WE      D S S L L N N Q F G T M P S L T M A C M A K Q S Q T P L N D V V Q A L T D L G L L Y T V K Y P N L S D L E R L
PV      D N T L L I N Q F G S M P A L T I A C M T E Q G G E Q L N D V V Q A L S A L G L L Y T V K F P N M T D L E K L
Common  D     L L   N Q F G   M P   L T   A C M     Q         L N D V V Q A L     L G L L Y T V K   P N     D L E   L

        221     225       230       235       240       245       250       255       260       265       270       275
WE      K D K H P V L G V I T E Q Q S S I N I S G Y N F S L G A A V K A G A A L L H G G N M L E S I L I K P S N S E D
PV      T Q Q H S A L K I I S H E P S A L N I S G Y N L S L S A A V K A A A C M I D G G N M L E T I Q V K P S M F S T
Common        H     L     I         S     N I S G Y N   S L   A A V K A   A         G G N M L E   I       K P S

        276     280       285       290       295       300       305       310       315       320       325       330
WE      L L K A V L G A K K K L N M F V S D Q V G D R N P Y E N I L Y K V C L S G E G W P Y I A C R T S V V G R A W E
PV      L I K S L L Q I K N R E G M F V S T T P G Q R N P Y E N L L Y K I C L S G D G W P Y I G S R S Q V Q G R A W D
Common  L   K     L     K       M F V S       G   R N P Y E N   L Y K   C L S G   G W P Y I     R     V   G R A W

        331     335       340       345       350       355       360       365       370       375       380       385
WE      N T T I D L T N E K L V A N S S R P V P G A A G P P Q V G L S Y S Q T M L L K D L M G G I D P N A P T W I D I
PV      N T T V D L D S K P S A I Q P P V R N G G S P D L K Q I P K E K E D T V V S S I Q M - - L D P R A T T W I D I
Common  N T T   D L                             G           Q             T           M       D P   A   T W I D I

        386     390       395       400       405       410       415       420       425       430       435       440
WE      E G R F N D P V E I A I F Q P Q N G Q F I H F Y R E P T D Q K Q F K Q D S K Y S H G M D L A D L F N A Q A G L
PV      E G T P N D P V E M A I Y Q P D T G N Y I H C Y R F P H D E K S F K E Q S K Y S H G L L L K D L A D A Q P G L
Common  E G     N D P V E   A I   Q P     I H   Y R   P   D   K   F K     S K Y S H G     L   D L     A Q   G L

        441     445       450       455       460       465       470       475       480       485       490       495
WE      T S S V I G A L P Q G M V L S C Q G S D D I R K L L D S Q N R R D I K L I D V E M T K E A S R E Y E D K V W D
PV      I S S I I R H L P Q N M V F T A Q G S D D I I R L F E M H G R R D L K V L D V K L S A E Q A R T F E D E I W E
Common    S S   I     L P Q   M V       Q G S D D I     L           R R D   K     D V         E     R     E D     W

        496     500       505       510       515       520       525       530       535       540       545       550
WE      K Y G W L C K M H T G V V R D K K K K E I T - - - - - P H C A L M D C I I F E S A S K A R L P D L K T V H N I
PV      R Y N Q L C T K H K G L V I K K K K K G A V Q T T A N P H C A L L D T I M F D A T V T G W V R D Q K P M R - C
Common    Y     L C     H   G   V     K K K K                 P H C A L   D   I   F                     D   K

        551     555       560       565
WE      L P H D L I F R G - P N V V T L
PV      L P I D T L Y R N N T D L I N L
Common  L P   D       R               L
```

A5 Nucleoprotein WE/LV

```
        1       5         10        15        20        25        30        35        40        45        50        55
WE      M S L S K E V K S F Q W T Q A L R R E L Q G F T S D V K A A V I K D A T S L L N G L D F S E V S N V Q R I M R
LV      M S A S K E V R S F L W T Q S L R R E L S G Y C S N I K L Q V V K D A Q A L L H G L D F S E V S N V Q R L M R
Common  M S   S K E V   S F     W T Q   L R R E L   G     S     K     V   K D A     L L   G L D F S E V S N V Q R   M R

        56      60        65        70        75        80        85        90        95        100       105       110
WE      K E R R D D K D L Q R L R S L N Q T V H S L V D P K S T S K K N V L K V G R L S A E E L M T L A A D L E K L K
LV      K Q K R D D G D L K R L R D L N Q A V N N L V E L K S T Q Q K S V L R V G T L S S D D L L I L A A D L E K L K
Common  K     R D D   D L   R L R   L N Q   V     L V     K S T     K   V L   V G   L S       L     L A A D L E K L K

        111     115       120       125       130       135       140       145       150       155       160       165
WE      A K I M R T E R P Q A S G V Y M G N L T A Q Q L D Q R S Q I L Q M V G M R R P Q Q G A S G - - - - V V R V W D
LV      S K V T R T E R P L S S G V Y M G N L S S Q Q L D Q R R A L L N M I G M T G V S G G G K G A S D G I V R V W D
Common    K     R T E R P     S G V Y M G N L     Q Q L D Q R       L   M   G M             G     G           V R V W D

        166     170       175       180       185       190       195       200       205       210       215       220
WE      V K D S S L L N N Q F G T M P S L T M A C M A K Q S Q T P L N D V V Q A L T D L G L L Y T V K Y P N L S D L E
LV      V K N A E L L N N Q F G T M P S L T L A C L T K Q G Q V D L N D A V Q A L T D L G L I Y T A K Y P N S S D L D
Common  V K       L L N N Q F G T M P S L T   A C     K Q   Q     L N D   V Q A L T D L G L   Y T   K Y P N   S D L

        221     225       230       235       240       245       250       255       260       265       270       275
WE      R L K D K H P V L G V I T E Q Q S S I N I S G Y N F S L G A A V K A G A A L L H G G N M L E S I L I K P S N S
LV      R L S Q S H P I L N M I D T K K S S L N I S G Y N F S L G A A V K A G A C M L D G G N M L E T I K V S P Q T M
Common  R L       H P   L   I       S S   N I S G Y N F S L G A A V K A G A     L   G G N M L E   I     P

        276     280       285       290       295       300       305       310       315       320       325       330
WE      E D L L K A V L G A K K K L N M F V S D Q V G D R N P Y E N I L Y K V C L S G E G W P Y I A C R T S V V G R A
LV      D G I L K S I L K V K K S L G M F V S D T P G E R N P Y E N I L Y K I C L S G D G W P Y I A S R T S I V G R A
Common        L K     L     K K   L   M F V S D     G   R N P Y E N I L Y K   C L S G   G W P Y I A   R T S   V G R A

        331     335       340       345       350       355       360       365       370       375       380       385
WE      W E N T T I D L T N E K L - - - V A N S S R P V P G A A G P P Q V G L S Y S Q T M L L K D L M G - G I D P N A
LV      W E N T V V D L E Q D N K P Q K I G N G G S N K S L Q S A G F A A G L T Y S Q L M T L K D F K C F N L I P N A
Common  W E N T   D L                   N                         G L   Y S Q   M   L K D             P N A

        386     390       395       400       405       410       415       420       425       430       435       440
WE      P T W I D I E G R F N D P V E I A I F Q P Q N G Q F I H F Y R E P T D Q K Q F K Q D S K Y S H G M D L A D L F
LV      K T W M D I E G R P E D P V E I A L Y Q P S S G C Y V H F F R E P T D L K Q F K Q D A K Y S H G I D V T D L F
Common    T W   D I E G R     D P V E I A     Q P     G     H F   R E P T D   K Q F K Q D   K Y S H G   D     D L F

        441     445       450       455       460       465       470       475       480       485       490       495
WE      N A Q A G L T S S V I G A L P Q G M V L S C Q G S D D I R K L L D S Q N R R D I K L I D V E M T K E A S R E Y
LV      A A Q P G L T S A V I E A L P R N M V I T C Q G S E D I R K L L E S Q G R R D I K L I D I T L S K A D S R K F
Common    A Q   G L T S   V I   A L P     M V     C Q G S   D I R K L L   S Q   R R D I K L I D         K     S R

        496     500       505       510       515       520       525       530       535       540       545       550
WE      E D K V W D K Y G W L C K M H T G V V R D K K K K - - - - E I T P H C A L M D C I I F E S A S K A R L P D L K
LV      E N A V W D Q F K D L C H M H T G V V V E K K K R G G K E E I T P H C A L M D C I M F D A A V S G G L - D A K
Common  E     V W D         L C   M H T G V V     K K K         E I T P H C A L M D C I   F     A         L   D   K

        551     555       560       565       570       575
WE      T V H N I L P H D L I F R G - - P N V V T L
LV      V L R V V L P R D M V F R T S T P K V V L -
Common            L P   D     F R     P . V V
```

A6 Nucleoprotein PV/LV

```
       1       5         10        15        20        25        30        35        40        45        50        55
PV     - - M S D N I P S F R W V Q S L T R G L S N W T H P V K A D V L S D T R A L L S A L D F H K V A Q V Q R M V R
LV     M S A S K E V R S F L W T Q S L R R E L S G Y C S N I K L Q V V K D A Q A L L H G L D F S E V S N V Q R L M R
Common       S         S F   W   Q S L   R   L S           K     V     D     A L L     L D F     V     V Q R     R

       56      60        65        70        75        80        85        90        95        100       105       110
PV     K D K R T D S D L T K L R D M N K E V D A L M N M R S V Q R D N V L K V G G L A K E E L M E L A S D L D K L R
LV     K Q K R D D G D L K R L R D L N Q A V N N L V E L K S T Q Q K S V L R V G T L S S D D L L I L A A D L E K L K
Common K   K R   D   D L     L R D   N     V     L         S   Q       V L   V G   L         L     L A   D L   K L

       111     115       120       125       130       135       140       145       150       155       160       165
PV     K K V T R T E G L S Q P G V Y E G N L T N T Q L E Q R A E I L R S M G F A N A R P A G N - - R D G V V K V W D
LV     S K V T R T E R P L S S G V Y M G N L S S Q Q L D Q R R A L L N M I G M T G V S G G G K G A S D G I V R V W D
Common   K V T R T E           G V Y   G N L       Q L   Q R       L       G             G         D G   V   V W D

       166     170       175       180       185       190       195       200       205       210       215       220
PV     I K D N T L L I N Q F G S M P A L T I A C M T E Q G G E Q L N D V V Q A L S A L G L L Y T V K F P N M T D L E
LV     V K N A E L L N N Q F G T M P S L T L A C L T K Q G Q V D L N D A V Q A L T D L G L I Y T A K Y P N S S D L D
Common   K       L L   N Q F G   M P   L T   A C   T   Q G       L N D   V Q A L     L G L   Y T   K   P N     D L

       221     225       230       235       240       245       250       255       260       265       270       275
PV     K L T Q Q H S A L K I I S H E P S A L N I S G Y N L S L S A A V K A A A C M I D G G N M L E T I Q V K P S M F
LV     R L S Q S H P I L N M I D T K K S S L N I S G Y N F S L G A A V K A G A C M L D G G N M L E T I K V S P Q T M
Common   L   Q   H       L   I         S   L N I S G Y N   S L   A A V K A   A C M   D G G N M L E T I   V   P

       276     280       285       290       295       300       305       310       315       320       325       330
PV     S T L I K S L L Q I K N R E G M F V S T T P G Q R N P Y E N L L Y K I C L S G D G W P Y I G S R S Q V Q G R A
LV     D G I L K S I L K V K K S L G M F V S D T P G E R N P Y E N I L Y K I C L S G D G W P Y I A S R T S I V G R A
Common         K S   L     K       G M F V S   T P G   R N P Y E N   L Y K I C L S G D G W P Y I   S R         G R A

       331     335       340       345       350       355       360       365       370       375       380       385
PV     W D N T T V D L D S K P S A I Q P P V R N G G S P D L K Q I P K E K E D T V V S S I Q M L D - - - - - - - - - -
LV     W E N T V V D L E Q D - - - - N K P Q K I G N G G S N K S L Q S A G F A A G L T Y S Q L M T L K D F K C F N L
Common W   N T   V D L                 P       G           K                             Q

       386     390       395       400       405       410       415       420       425       430       435       440
PV     - P R A T T W I D I E G T P N D P V E M A I Y Q P D T G N Y I H C Y R F P H D E K S F K E Q S K Y S H G L L L
LV     I P N A K T W M D I E G R P E D P V E I A L Y Q P S S G C Y V H F F R E P T D L K Q F K Q D A K Y S H G I D V
Common   P   A   T W   D I E G   P   D P V E   A   Y Q P     G   Y   H     R   P   D   K   F K     K Y S H G

       441     445       450       455       460       465       470       475       480       485       490       495
PV     K D L A D A Q P G L I S S I I R H L P Q N M V F T A Q G S D D I I R L F E M H G R R D L K V L D V K L S A E Q
LV     T D L F A A Q P G L T S A V I E A L P R N M V I T C Q G S E D I R K L L E S Q G R R D I K L I D I T L S K A D
Common   D L       A Q P G L   S       I       L P   N M V   T   Q G S   D I       L   E       G R R D   K       D       L S

       496     500       505       510       515       520       525       530       535       540       545       550
PV     A R T F E D E I W E R Y N Q L C T K H K G L V I K K K K K G A V Q T T A N P H C A L L D T I M F D A T V T G W
LV     S R K F E N A V W D Q F K D L C H M H T G V V V E K K K R G G K E E I T - P H C A L M D C I M F D A A V S G -
Common   R   F E       W           L C     H   G   V       K K K   G                     P H C A L   D   I M F D A   V   G

       551     555       560       565       570       575       580
PV     V R D Q K P M R C - L P I D T L Y R N N T D L I N L
LV     G L D A K V L R V V L P R D M V F R T S T P K V V L
Common     D   K       R         L P   D           R         T                   L
```

A7 Glycoprotein ARM/WE

```
1                                                             55
ARM     MGQIVTMFEALPHIIDEVINIVIIVLIVITGIKAVYNFATCGIFALISFLLLAGR
WE      MGQIVTMFEALPHIIDEVINIVIIVLIIITSIKAVYNFATCGILALVSFLFLAGR
Common  MGQIVTMFEALPHIIDEVINIVIIVLI IT IKAVYNFATCGI AL SFL LAGR

56                                                           110
ARM     SCGMYGLKGPDIYKGVYQFKSVEFDMSHLNLTMPNACSANNSHHYISMGTSGLEL
WE      SCGMYGLNGPDIYKGVYQFKSVEFDMSHLNLTMPNACSVNNSHHYISMGSSGLEP
Common  SCGMYGL GPDIYKGVYQFKSVEFDMSHLNLTMPNACS NNSHHYISMG SGLE

111                                                          165
ARM     TFTNDSIISHNFCNLTSAFNKKTFDHTLMSIVSSLHLSIRGNSNYKAVSCDFNNG
WE      TFTNDSILNHNFCNLTSALNKKSFDHTLMSIVSSLHLSIRGNSNYKAVSCDFNNG
Common  TFTNDSI  HNFCNLTSA NKK FDHTLMSIVSSLHLSIRGNSNYKAVSCDFNNG

166                                                          220
ARM     ITIQYNLTFSDAQSAQSQCRTFRGRVLDMFRTAFGGKYMRSGWGWTGSDGKTTWC
WE      ITIQYNLSSSDPQSAMSQCRTFRGRVLDMFRTAFGGKYMRSGWGWTGSDGYTTWC
Common  ITIQYNL  SD QSA SQCRTFRGRVLDMFRTAFGGKYMRSGWGWTGSDG TTWC

221                                                          275
ARM     SQTSYQYLIIQNRTWENHCTYAGPFGMSRILLSQEKTKFFTRRLAGTFTWTLSDS
WE      SQTSYQYLIIQNRTWENHCRYAGPFGMSRILFAQEKTKFLTRRLSGTFTWTLSDS
Common  SQTSYQYLIIQNRTWENHC YAGPFGMSRIL  QEKTKF TRRL GTFTWTLSDS

276                                                          330
ARM     SGVENPGGYCLTKWMILAAELKCFGNTAVAKCNVNHDAEFCDMLRLIDYNKAALS
WE      SGVENPGGYCLTKWMILAAELKCFGNTAVAKCNVNHDEEFCDMLRLIDYNKAALS
Common  SGVENPGGYCLTKWMILAAELKCFGNTAVAKCNVNHD EFCDMLRLIDYNKAALS

331                                                          385
ARM     KFKEDVESALHLFKTTVNSLISDQLLMRNHLRDLMGVPYCNYSKFWYLEHAKTGE
WE      KFKQDVESALHVFKTTLNSLISDQLLMRNHLRDLMGVPYCNYSKFWYLEHAKTGE
Common  KFK DVESALH FKTT NSLISDQLLMRNHLRDLMGVPYCNYSKFWYLEHAKTGE

386                                                          440
ARM     TSVPKCWLVTNGSYLNETHFSDQIEQEADNMITEMLRKDYIKRQGSTPLALMDLL
WE      TSVPKCWLVTNGSYLNEIHFSDQIEQEADNMITEMLRKDYIKRQGSTPLALMDLL
Common  TSVPKCWLVTNGSYLNE HFSDQIEQEADNMITEMLRKDYIKRQGSTPLALMDLL

441                                                          495
ARM     MFSTSAYLVSIFLHLVKIPTHRHIKGGSCPKPHRLTNKGICSCGAFKVPGVKTVW
WE      MFSTSAYLISIFLHFVRIPTHRHIKGGSCPKPHRLTNKGICSCGAFKVPGVKTIW
Common  MFSTSAYL SIFLH V IPTHRHIKGGSCPKPHRLTNKGICSCGAFKVPGVKT W

496   500
ARM     KRR
WE      KRR
Common  KRR
```

A8 Glycoprotein ARM/PV

```
         1         5         10        15        20        25        30        35        40        45        50        55
ARM      M G Q I V T M F E A L P H I I D E V I N I V I I V L I V I T G I K A V Y N F A T C G I F A L I S F L L L A G R
PV       M G Q I V T L I Q S I P E V L Q E V F N V A L I I V S V L C I V K G F V N L M R C G L F Q L V T F L I L S G R
Common   M G Q I V T           P         E V   N       I       V         K       N       C G   F   L       F L   L   G R

         56      60        65        70        75        80        85        90        95        100       105       110
ARM      S C G M Y G L K G P D I Y K G V Y Q F K S V E F D M S H L N L T M P N A C S A N N S H H Y I - - - S M G T S G
PV       S C D - - - - - - S M M I D R R H N L T H V E F N L T R M F D N L P Q S C S K N N T H H Y Y K G P S N T T W G
Common   S C                                           V E F                 P     C S   N N   H H Y         S     T   G

         111     115       120       125       130       135       140       145       150       155       160       165
ARM      L E L T F T N D S I I S H - - - N F C N L T S - A F N K - - - - - - - K T F D H T L M S I V S S L H L S I R G
PV       I E L T L T N T S I A N E T S G N F S N I G S L G Y G N I S N C D R T R E A G H T L K W L L N E L H F N V L H
Common     E L T   T N   S I             N F   N       S                                 H T L             L H

         166     170       175       180       185       190       195       200       205       210       215       220
ARM      N S N Y K A V S C D F - - N N G I T I Q Y N L T F S D A Q S A Q S Q C R T F R G R V L D M F - - - R T A F G G
PV       V T R H I G A R C K T V E G A G V L I Q Y N L T V G D R G G E V G - - R H L I A S L A Q I I G D P K I A W V G
Common                   C             G       I Q Y N L T     D                 R                               A     G

         221     225       230       235       240       245       250       255       260       265       270       275
ARM      K Y M R S G W G W T G S D G K T T W C - S Q T S Y Q Y L I I Q N R T W E N H C T Y A G P F G M S R I L L S Q E
PV       K C F N N C - - - S G D T C R L T N C E G G T H Y N F L I I Q N T T W E N H C T Y T P - - - M A T I R M A L Q
Common   K                   G           T   C       T   Y     L I I Q N   T W E N H C T Y           M     I

         276     280       285       290       295       300       305       310       315       320       325       330
ARM      K T K F - - F T R R L A G T F T W T L S D S S G V E N P G G Y C L T K W M I L A A E L K C F G N T A V A K C N
PV       R T A Y S S V S R K L L G F F T W D L S D S S G Q H V P G G Y C L E Q W A I I W A G I K C F D N T V M A K C N
Common     T             R   L   G   F T W   L S D S S G       P G G Y C L     W   I     A     K C F   N T     A K C N

         331     335       340       345       350       355       360       365       370       375       380       385
ARM      V N H D A E F C D M L R L I D Y N K A A L S K F K E D V E S A L H L F K T T V N S L I S D Q L L M R N H L R D
PV       K D H N E E F C D T M R L F D F N Q N A I K T L Q L N V E N S L N L F K K T I N G L I S D S L V I R N S L K Q
Common       H     E F C D     R L   D   N     A               V E     L   L F K   T   N   L I S D   L     R N   L

         386     390       395       400       405       410       415       420       425       430       435       440
ARM      L M G V P Y C N Y S K F W Y L E H A K T G E T S V P K C W L V T N G S Y L N E T H F S D Q I E Q E A D N M I T
PV       L A K I P Y C N Y T K F W Y I N D T I T G R H S L R Q C W L V H N G S Y L N E T H F K N D W L W E S Q N L Y N
Common   L       P Y C N Y   K F W Y           T G     S   P   C W L V   N G S Y L N E T H F             E     N

         441     445       450       455       460       465       470       475       480       485       490       495
ARM      E M L R K D Y I K R Q G S T P L A L M D L - - - - L M F S T S A Y L V S I F L H L V K I P T H R H I K G G S C
PV       E M L M K E Y E E R Q G K T P L A L T D I C F W S L V F - - - - Y T I T V F L H I V G I P T H R H I I G D G C
Common   E M L   K   Y     R Q G   T P L A L   D           L   F         Y         F L H   I   P T H R H I   G     C

         496     500       505       510       515       520       525       530
ARM      P K P H R L T N K G I C S C G A F K V P G V K T V W K R R - - - -
PV       P K P H R I T R N S L C S C G Y Y K - - - - - - - Y Q R N L T N G
Common   P K P H R   T         C S C G     K                   R
```

A9 Glycoprotein ARM/LV

```
1                                                                    55
ARM:    MGQIVTMFEALPHIIDEVINIVIIVLIVITGIKAVYNFATCGIFALISFLLLAGR
LV:     MGQIVTFFQEVPHVIEEVMNIVLIALSVLAVLKGLYNFATCGLVGLVTFLLLCGR
Common: MGQIVT F   PH I EV NIV I L V    K  YNFATCG   L  FLLL GR

56                                                                  110
ARM:    SCGMYGLKGPDIYKGVYQFKSVEFDMSHLNLTMPNACSANNSHHYISMGT-SGLE
LV:     SCT------TSLYKGVYELQTLELNMETLNMTMPLSCTKNNSHHYIMVGNETGLE
Common: SC          YKGVY     E  M  LN TMP  C  NNSHHYI  G   GLE

111                                                                 165
ARM:    LTFTNDSIISHNFCNLTSAFNKKTFDHTLMSIVSSLHLSIRGNSNYKAVSCDFNN
LV:     LTLTNTSIINHKFCNLSDAHKKNLYDHALMSIISTFHLSIPNFNQYEAMSCDFNG
Common: LT TN SII H FCNL  A  K   DH LMSI S  HLSI     Y A SCDFN

166                                                                 220
ARM:    G-ITIQYNLTFSDAQSAQSQCRTFRGRVLDMF-RTAFGGKYMRSGWGWTGSDGKT
LV:     GKISVQYNLSHSYAGDAANHCGTVANGVLQTFMRMAWGGSYI------ALDSGRG
Common: G I  QYNL  S A  A   C T    VL F R A GG Y           G

221                                                                 275
ARM:    TW-CSQTSYQYLIIQNRTWENHCTYA--GPFGMSRILLSQEKTKFFTRRLAGTFT
LV:     NWDCIMTSYQYLIIQNTTWEDHCQFSRPSPIGYLGLLSQRTRDIYISRRLLGTFT
Common:  W C  TSYQYLIIQN TWE HC      P G         L          RRL GTFT

276                                                                 330
ARM:    WTLSDSSGVENPGGYCLTKWMILAAELKCFGNTAVAKCNVNHDAEFCDMLRLIDY
LV:     WTLSDSEGKDTPGGYCLTRWMLIEAELKCFGNTAVAKCNEKHDEEFCDMLRLFDF
Common: WTLSDS G   PGGYCLT WM   AELKCFGNTAVAKCN  HD EFCDMLRL D

331                                                                 385
ARM:    NKAALSKFKEDVESALHLFKTTVNSLISDQLLMRNHLRDLMGVPYCNYSKFWYLE
LV:     NKQAIQRLKAEAQMSIQLINKAVNALINDQLIMKNHLRDIMGIPYCNYSKYWYLN
Common: NK A  K K        L    VN LI DQL M NHLRD MG PYCNYSK WYL

386                                                                 440
ARM:    HAKTGETSVPKCWLVTNGSYLNETHFSDQIEQEADNMITEMLRKDYIKRQGSTPL
LV:     HTTTGRTSLPKCWLVSNGSYLNETHFSDDIEQQADNMITEMLQKEYMERQGKTPL
Common: H  TG TS PKCWLV NGSYLNETHFSD IEQ ADNMITEML K Y  RQG TPL

441                                                                 495
ARM:    ALMDLLMFSTSAYLVSIFLHLVKIPTHRHIKGGSCPKPHRLTNKGICSCGAFKVP
LV:     GLVDLFVFSTSFYLISIFLHLVKIPTHRHIVGKSCPKPHRLNHMGICSCGLYKQP
Common:  L DL  FSTS YL SIFLHLVKIPTHRHI G SCPKPHRL   GICSCG  K P

496      505
ARM:    GVKTVWKRR
LV:     GVPVKWKR-
Common: GV   WKR
```

A10 Glycoprotein WE/PV

```
            1         5         10        15        20        25        30        35        40        45        50        55
WE          M G Q I V T M F E A L P H I I D E V I N I V I I V L I I I T S I K A V Y N F A T C G I L A L V S F L F L A G R
PV          M G Q I V T L I Q S I P E V L Q E V F N V A L I I V S V L C I V K G F V N L M R C G L F Q L V T F L I L S G R
Common      M G Q I V T           P         E V   N     I                 K       N     C G         L V   F L   L   G R

            56        60        65        70        75        80        85        90        95        100       105       110
WE          S C G M Y G L N G P D I Y K G V Y Q F K S V E F D M S H L N L T M P N A C S V N N S H H Y I S M G S S - - - G
PV          S C D S M M I D R R H - - - - - - N L T H V E F N L T R M F D N L P Q S C S K N N T H H Y Y K G P S N T T W G
Common      S C                             V E F             P       C S     N N     H H Y               S         G

            111       115       120       125       130       135       140       145       150       155       160
WE          L E P T F T N D S I L N H N - - - - - - - - - - - F C N L T S A L N K K S F D H T L M S I V S S L H L S I R G
PV          I E L T L T N T S I A N E T S G N F S N I G S L G Y G N I S N C D R T R E A G H T L K W L L N E L H F N V L H
Common        E     T     T N   S I   N                         N                       H T L             L H

            166       170       175       180       185       190       195       200       205       210       215       220
WE          N S N Y K A V S C D F - - N N G I T I Q Y N L S S S D P Q S A M S Q C R T F R G R V L D M F - - - R T A F G G
PV          V T R H I G A R C K T V E G A G V L I Q Y N L T V G D R G G E V G - - R H L I A S L A Q I I G D P K I A W V G
Common                      C           G       I Q Y N L       D           R                               A       G

            221       225       230       235       240       245       250       255       260       265       270       275
WE          K Y M R S G W G W T G S D G Y T T W C - S Q T S Y Q Y L I I Q N R T W E N H C R Y A G P F G M S R I L F A Q E
PV          K C F N N C - - - S G D T C R L T N C E G G T H Y N F L I I Q N T T W E N H C T Y T P - - - M A T I R M A L Q
Common      K                 G           T   C           T   Y     L I I Q N     T W E N H C   Y             M     I       A

            276       280       285       290       295       300       305       310       315       320       325       330
WE          K T K F - - L T R R L S G T F T W T L S D S S G V E N P G G Y C L T K W M I L A A E L K C F G N T A V A K C N
PV          R T A Y S S V S R K L L G F F T W D L S D S S G Q H V P G G Y C L E Q W A I I W A G I K C F D N T V M A K C N
Common        T                 R   L   G     F T W   L S D S S G         P G G Y C L     W     I           K C F   N T       A K C N

            331       335       340       345       350       355       360       365       370       375       380       385
WE          V N H D E E F C D M L R L I D Y N K A A L S K F K Q D V E S A L H V F K T T L N S L I S D Q L L M R N H L R D
PV          K D H N E E F C D T M R L F D F N Q N A I K T L Q L N V E N S L N L F K K T I N G L I S D S L V I R N S L K Q
Common            H   E E F C D     R L   D   N       A               V E     L         F K   T   N       L I S D   L       R N       L

            386       390       395       400       405       410       415       420       425       430       435
WE          L M G V P Y C N Y S K F W Y L E H A K T G E T S V R K C W L V T N G S Y L N E I H F S D Q I E Q E A D N M I T
PV          L A K I P Y C N Y T K F W Y I N D T I T G R H S L P Q C W L V H N G S Y L N E T H F K N D W L W E S Q N L Y N
Common      L         P Y C N Y     K F W Y             T G       S     P     C W L V     N G S Y L N E     H F                         E                       N

            441       445       450       455       460       465       470       475       480       485       490       495
WE          E M L R K D Y I K R Q G S T P L A L M D L - - - - L M F S T S A Y L I S I F L H F V R I P T H R H I K G G S C
PV          E M L M K E Y E E R Q G K T P L A L T D I C F W S L V F - - - - Y T I T V F L H I V G I P T H R H I I G D G C
Common      E M L   K   Y       R Q G   T P L A L   D             L   F                 Y   I       F L H   V     I P T H R H I       G           C

            496       500       505       510       515       520       525       530
WE          P K P H R L T N K G I C S C G A F K V P G V K T I W K R R - - - -
PV          P K P H R I T R N S L C S C G Y Y K - - - - - - - - Y Q R N L T N G
Common      P K P H R   T           C S C G       K                     R
```

A11 Glycoprotein WE/LV

```
        1       5         10        15        20        25        30        35        40        45        50        55
WE      M G Q I V T M F E A L P H I I D E V I N I V I I V L I I I T S I K A V Y N F A T C G I L A L V S F L F L A G R
LV      M G Q I V T F F Q E V P H V I E E V M N I V L I A L S V L A V L K G L Y N F A T C G L V G L V T F L L L C G R
Common  M G Q I V T   F       P H   I   E V   N I V   I   L             K     Y N F A T C G       L V   F L   L   G R

        56      60        65        70        75        80        85        90        95        100     105     110
WE      S C G M Y G L N G P D I Y K G V Y Q F K S V E F D M S H L N L T M P N A C S V N N S H H Y I S M G S - S G L E
LV      S C T - - - - - - T S L Y K G V Y E L Q T L E L N M E T L N M T M P L S C T K N N S H H Y I M V G N E T G L E
Common  S C                       Y K G V Y           E     M     L N   T M P     C     N N S H H Y I       G       G L E

        111     115       120       125       130       135       140       145       150       155       160       165
WE      P T F T N D S I L N H N F C N L T S A L N K K S F D H T L M S I V S S L H L S I R G N S N Y K A V S C D F N N
LV      L T L T N T S I I N H K F C N L S D A H K K N L Y D H A L M S I I S T F H L S I P N F N Q Y E A M S C D F N G
Common    T   T N   S I   N H   F C N L     A     K       D H   L M S I   S     H L S I         Y   A   S C D F N

        166     170       175       180       185       190       195       200       205       210       215       220
WE      G - I T I Q Y N L S S S D P Q S A M S Q C R T F R G R V L D M F - R T A F G G K Y - - M R S G W G - W T G S D
LV      G K I S V Q Y N L S H S Y A G D A A N H C G T V A N G V L Q T F M R M A W G G S Y I A L D S G R G N W D - - -
Common  G   I     Q Y N L S   S         A       C   T         V L     F   R   A   G G   Y         S G   G   W

        221     225       230       235       240       245       250       255       260       265       270       275
WE      G Y T T W C S Q T S Y Q Y L I I Q N R T W E N H C R Y A - - G P F G M S R I L F A Q E K T K F L T R R L S G T
LV      - - - - - C I M T S Y Q Y L I I Q N T T W E D H C Q F S R P S P I G Y L G L L S Q R T R D I Y I S R R L L G T
Common            C     T S Y Q Y L I I Q N   T W E   H C             P   G         L                       R R L   G T

        276     280       285       290       295       300       305       310       315       320       325       330
WE      F T W T L S D S S G V E N P G G Y C L T K W M I L A A E L K C F G N T A V A K C N V N H D E E F C D M L R L I
LV      F T W T L S D S E G K D T P G G Y C L T R W M L I E A E L K C F G N T A V A K C N E K H D E E F C D M L R L F
Common  F T W T L S D S   G       P G G Y C L T   W M       A E L K C F G N T A V A K C N     H D E E F C D M L R L

        331     335       340       345       350       355       360       365       370       375       380       385
WE      D Y N K A A L S K F K Q D V E S A L H V F K T T L N S L I S D Q L L M R N H L R D L M G V P Y C N Y S K F W Y
LV      D F N K Q A I Q R L K A E A Q M S I Q L I N K A V N A L I N D Q L I M K N H L R D I M G I P Y C N Y S K Y W Y
Common  D   N K   A                               N   L I   D Q L   M   N H L R D   M G   P Y C N Y S K   W Y

        386     390       395       400       405       410       415       420       425       430       435       440
WE      L E H A K T G E T S V P K C W L V T N G S Y L N E I H F S D Q I E Q E A D N M I T E M L R K D Y I K R Q G S T
LV      L N H T T T G R T S L P K C W L V S N G S Y L N E T H F S D D I E Q Q A D N M I T E M L Q K E Y M E R Q G K T
Common  L   H     T G   T S   P K C W L V   N G S Y L N E   H F S D   I E Q   A D N M I T E M L   K   Y     R Q G   T

        441     445       450       455       460       465       470       475       480       485       490       495
WE      P L A L M D L L M F S T S A Y L I S I F L H F V R I P T H R H I K G G S C P K P H R L T N K G I C S C G A F K
LV      P L G L V D L F V F S T S F Y L I S I F L H L V K I P T H R H I V G K S C P K P H R L N H M G I C S C G L Y K
Common  P L   L   D L     F S T S   Y L I S I F L H   V   I P T H R H I   G   S C P K P H R L       G I C S C G   K

        496     500       505       510
WE      V P G V K T I W K R R
LV      Q P G V P V K W K R -
Common    P G V       W K R
```

A12 Glycoprotein PV/LV

```
        1       5         10        15        20        25        30        35        40        45        50        55
PV      M G Q I V T L I Q S I P E V L Q E V F N V A L I I V S V L C I V K G F V N L M R C G L F Q L V T F L I L S G R
LV      M G Q I V T F F Q E V P H V I E E V M N I V L I A L S V L A V L K G L Y N F A T C G L V G L V T F L L L C G R
Common  M G Q I V T     Q     P   V     E V   N     L I     S V L       K G     N         C G L     L V T F L   L   G R

        56      60        65        70        75        80        85        90        95        100       105       110
PV      S C D S M M I D R R H N L T H V E F N L T R M F D N L P Q S C S K N N T H H Y Y K G P S N T T W G I E L T L T
LV      S C T T S L Y K G V Y E L Q T L E L N M E T L N M T M P L S C T K N N S H H Y I M V G N E T - - G L E L T L T
Common  S C                     L       E   N                 P   S C   K N N   H H Y             T     G   E L T L T

        111     115       120       125       130       135       140       145       150       155       160       165
PV      N T S I A N E T S G N F S N I G S L G Y G N I S N C D R T R E A G H T L K W L L N E L H F N V L H V T R H I G
LV      N T S I I N H K - - - - - - - - - - - F C N L S D A H K K N L Y D H A L M S I I S T F H L S I P N F N Q Y E A
Common  N T S I   N                                 N   S                   H   L                 H

        166     170       175       180       185       190       195       200       205       210       215       220
PV      A R C K T V E G A G V L I Q Y N L T - - - V G D R G G E V G R H L I A S L A Q I I G D P K I A W V G K C - - F
LV      M S C D - F N G G K I S V Q Y N L S H S Y A G D A A N H C G T V A N G V L Q T F M - - - R M A W G G S Y I A L
Common      C         G           Q Y N L           G D                 G                         L                 A W     G

        221     225       230       235       240       245       250       255       260       265       270       275
PV      N N C S G D - T C R L T N C E G G T H Y N F L I I Q N T T W E N H C T Y T - - - P M A T I R M A L Q R T A Y S
LV      D S G R G N W D C I M T S - - - - - - Y Q Y L I I Q N T T W E D H C Q F S R P S P I G Y L G L L S Q R T R D I
Common          G       C       T                 Y     L I I Q N T T W E   H C                       P                 Q R T

        276     280       285       290       295       300       305       310       315       320       325
PV      S V S R K L L G F F T W D L S D S S G Q H V P G G Y C L E Q W A I I W A G I K C F D N T V M A K C N K D H N E
LV      Y I S R R L L G T F T W T L S D S E G K D T P G G Y C L T R W M L I E A E L K C F G N T A V A K C N E K H D E
Common      S R   L L G   F T W   L S D S   G       P G G Y C L     W     I   A     K C F   N T     A K C N     H   E

        331     335       340       345       350       355       360       365       370       375       380       385
PV      E F C D T M R L F D F N Q N A I K T L Q L N V E N S L N L F K K T I N G L I S D S L V I R N S L K Q L A K I P
LV      E F C D M L R L F D F N K Q A I Q R L K A E A Q M S I Q L I N K A V N A L I N D Q L I M K N H L R D I M G I P
Common  E F C D     R L F D F N     A I     L           S     L     K     N   L I   D   L         N   L         I P

        386     390       395       400       405       410       415       420       425       430       435       440
PV      Y C N Y T K F W Y I N D T I T G R H S L P Q C W L V H N G S Y L N E T H F K N D W L W E S Q N L Y N E M L M K
LV      Y C N Y S K Y W Y L N H T T T G R T S L P K C W L V S N G S Y L N E T H F S D D I E Q Q A D N M I T E M L Q K
Common  Y C N Y   K   W Y   N   T   T G R   S L P   C W L V   N G S Y L N E T H F     D             N         E M L   K

        441     445       450       455       460       465       470       475       480       485       490
PV      E Y E E R Q G K T P L A L T D I C F W S L V F Y T I T V F L H I V G I P T H R H I I G D G C P K P H R I T R N
LV      E Y M E R Q G K T P L G L V D L F V F S T S F Y L I S I F L H L V K I P T H R H I V G K S C P K P H R L N H M
Common  E Y   E R Q G K T P L   L   D         S     F Y   I       F L H   V   I P T H R H I   G     C P K P H R

        496     500       505       510       515
PV      S L C S C G Y Y K Y Q R N L T N G - - -
LV      G I C S C G L Y K - Q P G V P V K W K R
Common      C S C G   Y K   Q
```

References

Ahmed R, Salmi A, Butler LD, Chiller JM, Oldstone MBA (1984) Selection of genetic variants of lymphocytic choriomeningitis virus in spleens of persistently infected mice. Role in suppression of cytotoxic T lymphocyte response and viral persistence. J Exp Med 60:521–540

Auperin DD, Compans RW, Bishop DHL (1982a) Nucleotide sequence conservation at the 3' termini of the virion RNA species of new world and old world arenaviruses. Virology 121:200–203

Auperin DD, Dimock K, Cash P, Rawls WE, Leung W-C, Bishop DHL (1982b) Analyses of the genomes of prototype Pichinde arenavirus and a virulent derivative Pichinde munchique: evidence for sequence conservation at the 3' termini of their viral RNA species. Virology 116:363–367

Auperin DD, Romanowski V, Galinsky M, Bishop DHL (1984) Sequencing studies of Pichinde arenavirus S RNA indicate a novel coding strategy, an ambisense viral S RNA. J Virol 52:897–904

Auperin DD, Sasso DR, McCormick JB (1986) Nucleotide sequence of the glycoprotein gene and intergenic region of the Lassa virus S genome RNA. Virology 154:155–167

Buchmeier MJ (1984) Antigenic and structural studies on the glycoproteins of lymphocytic choriomeningitis virus. In: Compans RW, Bishop DHL (eds) Segmented negative strand viruses. Academic, Orlando, pp 193–200

Buchmeier MJ, Lewicki HA, Tomori O, Johnson KM (1980) Monoclonal antibodies to lymphocytic choriomeningitis virus react with pathogenic arenaviruses. Nature 288:486–487

Buchmeier MJ, Lewicki HA, Tomori O, Oldstone MBA (1981) Monoclonal antibodies to lymphocytic choriomeningitis virus and Pichinde viruses: generation, characterization and cross-reactivity with other arenaviruses. Virology 113:73–85

Clegg JCS, Oram JD (1985) Molecular cloning of lassa virus RNA: nucleotide sequence and expression of the nucleocapsid protein gene. Virology 114:363–372

Domingo E, Martinez-Salas E, Sobrino F (1985) The quasispecies (extremely heterogeneous) nature of viral RNA genome populations: biological relevance – a review. Gene 40:1–8

Dutko F, Oldstone MBA (1983) Genomic and biological variation among commonly used lymphocytic choriomeningitis virus strains. J Gen Virol 64:1689–1698

Ghosh PK, Reddy VB, Piatak M, Lebowitz P, Weissman SM (1980) Determination of RNA sequences by primer directed synthesis and sequencing of their cDNA transcripts. Methods Enzymol 65:580–595

Gilbert W (1981) DNA sequencing and gene structure. Science 214:1305–1312

Gluzman Y (ed) (1985) Eukaryotic transcription. Cold Spring Harbor Laboratory Press, New York

Gluzman Y, Shenk T (eds) (1983) Enhancers and eukaryotic gene expression. Cold Spring Harbor Laboratory Press, New York

Holland JJ (1984) Continuum of change in RNA virus genomes. *In:* Notkins AL, Oldstone MBA (eds) Concepts in viral pathogenesis. Springer, Berlin Heidelberg New York Tokyo, pp 137–143

Kamer G, Argos P (1984) Primary structural comparison of RNA-dependent polymerases from plant, animal and bacterial viruses. Nucleic Acids Res 12:7269–7282

Lerner R (1982) Tapping the immunological repertoire to produce antibodies of predetermined specificity. Nature 299:592–596

Parekh BS, Buchmeier MJ (1986) Proteins of lymphocytic choriomeningitis virus: Antigenic topography of the viral glycoproteins. Virology 153:168–178

Pedersen IR (1979) Structural components and replication of arenaviruses. Adv Virus Res 24:277–330

Reanney DC (1982) The evolution of RNA viruses. Ann Rev Microbiol 36:47–83

Riviere Y, Ahmed R, Southern PJ, Buchmeier MJ, Oldstone MBA (1985a) Genetic mapping of lymphocytic choriomeningitis virus pathogenicity: virulence in guinea pigs is associated with the L RNA segment. J Virol 55:704–709

Riviere Y, Ahmed R, Southern PJ, Buchmeier MJ, Dutko FJ, Oldstone MBA (1985b) The S RNA segment of lymphocytic choriomeningitis virus codes for the nucleoprotein and glycoproteins 1 and 2. J Virol 53:966–968

Riviere Y, Ahmed R, Southern PJ, Oldstone MBA (1985c) Perturbation of differentiated functions during viral infection in vivo II. Viral reassortants map growth hormone defect to the S RNA of the lymphocytic choriomeningitis virus genome. Virology 142:175–182

Romanowski V, Bishop DHL (1985) Conserved sequences and coding of two strains of lymphocytic choriomeningitis virus (WE and Arm) and Pichinde arenavirus. Virus Res 2:35–51

Romanowski V, Matsuura Y, Bishop DHL (1985) Complete sequence of the S RNA of lymphocytic choriomeningitis virus (WE strain) compared to that of Pichinde arenavirus. Virus Res 3:101–114
Sanger F (1981) Determination of nucleotide sequences in DNA. Science 214:1205–1210
Schubert M, Harmison GG, Meier E (1984) Primary structure of the vesicular stomatitis virus polymerase (L) gene: evidence for a high frequency of mutations. J Virol 51:505–514
Southern PJ, Buchmeier MJ, Ahmed R, Francis SJ, Parekh B, Riviere Y, Singh MK, Oldstone MBA (1986a) Molecular pathogenesis of arenavirus infections. In: Brown F, Chanock RM, Lerner RA (eds) Vaccines 86. New approaches to immunization. Cold Spring Harbor Laboratory Press, New York, pp 239–245
Southern PJ, Singh MK, Riviere Y, Jacoby DR, Buchmeier MJ, Oldstone MBA (1986b) Molecular characterisation of the genomic S RNA segment from lymphocytic choriomeningitis virus. Virology
Wigler M, Levy D, Perucho M (1981) The somatic replication of DNA methylation. Cell 24:33–40

Protein Structure and Expression Among Arenaviruses

M.J. BUCHMEIER and B.S. PAREKH

1 Introduction

The proteins of arenaviruses were first studies by SMADEL and his colleagues (1939, 1940, 1942), with reference to their antigenicity. These workers described the presence of a virus-specific soluble (S) antigen detectable by complement fixation (CF) in homogenates of spleen and lung from LCMV-infected guinea pigs. Soluble antigen could be separated from infectious virus by ultracentrifugation. Repeatedly washed virions reacted poorly in CF tests while the S antigen lost none of its immunoreactivity after ultracentrifugation. These studies were not extended until nearly 3 decades later when BROWN and KIRK (1969), CHASTEL (1970), SIMON (1970), and BRO-JORGENSEN (1971) described antigens detectable by CF and immunodiffusion in tissues or cell cultures infected with LCMV. BRO-JORGENSEN (1971) found two antigenic species by immunodiffusion using infected BHK-21 cells as the antigen source. One antigen was heat stable and resistant to proteolysis, while the second was degraded by both heating and pronase digestion. Both antigens sedimented at a rate of 3.5 S in rate zonal sucrose gradient centrifugation, and based on this S value the molecular weight of the thermolabile S antigen was estimated to be 48 000.

Studies by GSCHWENDER (1976) with LCMV established that the extractable complement-fixing antigen (ECFA) was an internal component of the virion. Antiserum directed against ECFA did not neutralize infectious LCMV, and it did not mediate complement-dependent lysis of LCMV-infected cells. Purified LCMV disrupted by detergents liberated an antigen which reacted with anti-ECFA in CF test and produced a band of identity with ECFA by immunodiffusion.

Department of Immunology, Scripps Clinic and Research Foundation, 10666 North Torrey Pines Road, La Jolla, CA 92037, USA

Current Topics in Microbiology and Immunology, Vol. 133
© Springer-Verlag Berlin · Heidelberg 1987

RAWLS and BUCHMEIER (1975) arrived at similar conclusions, working with the S antigens of Pichinde virus. Antisera directed against partially purified CF antigen from cells infected by Pichinde virus were shown to react against the internal nucleocapsid protein of the virus but not with surface antigens of infected cells. Subsequent studies (BUCHMEIER et al. 1977) demonstrated that the S antigen was a degradation product of the viral nucleocapsid protein (NP). Moreover the antigenic cross-reactivity observed by CF among the new world, Tacaribe complex arenaviruses (reviewed in RAWLS and LEUNG 1979) was shown to be due to conservation of NP-related antigens (BUCHMEIER and OLDSTONE 1977).

Persistent arenavirus infections, whether in vitro (LEHMANN-GRUBE et al. 1969; WELSH and OLDSTONE 1977; WELSH and BUCHMEIER 1979; VAN DER ZEIJST et al. 1983a, b) or in vivo (TRAUB 1936; WILSNACK and ROWE 1964; OLDSTONE and BUCHMEIER 1982; RODRIGUEZ et al. 1983; see also FRANCIS et al., this volume), are characterized by the persistence of nucleoprotein, often in the absence of viral glycoprotein antigens and infectious virus production. This phenomenon, which we now recognize may be a consequence of the ambisense genomic arrangement of the S RNA segment (see BISHOP and AUPERIN, this volume), led to a great deal of confusion and ambiguity in early attempts to grow and purify arenaviruses (RAWLS and LEUNG 1979). Only when factors such as multiplicity of infection, rigorous plaque purification of the infecting virus, and time of harvest were carefully controlled did a consistent picture of the structural features of these viruses emerge.

2 Structural Proteins of Arenaviruses

The structural proteins of purified arenaviruses were first studied by RAMOS et al. (1972) working with Pichinde virus and by PEDERSON (1973) with LCMV. Numerous other descriptive studies of the proteins of these agents have followed, and to date the structural proteins of at least nine different arenaviruses have been examined (summarized in Table 1). Despite apparent differences, a number of common features have emerged. Characteristically, arenaviruses contain a major dominating protein which is the viral nucleocapsid protein (NP; 60–68 K). Nucleocapsid protein constitutes up to 58% of the protein in arenaviruses (VEZZA et al. 1977) and is easily detected in SDS-PAGE gels by protein staining or by radiolabeling with amino acid precursors such as [^{3}H]leucine or [^{35}S]methionine (Fig. 1). The viruses also contain variably either one, as reported for Tacaribe and Tamiami (GARD et al. 1977) or two, for LCMV and Pichinde virus (BUCHMEIER et al. 1978; VEZZA et al. 1977), glycopeptides of somewhat lower molecular weight than NP. Other minor proteins have also been detected but the origin of these has until recently been largely a subject of conjecture. Predominant among these quantitatively minor proteins are a 180–200 K protein termed L in Pichinde (HARNISH et al. 1981, 1983) and LCMV (BUCHMEIER, SINGH, and SOUTHERN, unpublished observation), which is a candidate for a viral polymerase, a 77 K protein termed P (VEZZA et al. 1977; GARD et al.

Table 1. Structural polypeptides of arenaviruses

Virus	L proteins		Nucleoproteins		Envelope proteins		Minor proteins		References
	Molecular weight (K)	Name	Molecular weight (K)	Name	Molecular weight (K)	Name	Molecular weight (K)	Name	
LCMV			63	NP	44	GP1	–	–	Buchmeier et al. 1978; Buchmeier and Oldstone 1979
					35	GP2			
LCMV	200	P 200	63	p 63	130	gp 130	77	p 77	Bruns et al. 1983b
					85	gp 85	38[a]	p 38	
					60	gp 60	26[a]	p 26	
					44	gp 44	25[a]	p 25	
					35	gp 35	19[a]	p 19	
Pichinde	200	L	64	NP	52	GP1	48[a]		Harnish et al. 1981
					36	GP2	38[b]		
							28[a]		
Pichinde	–	–	66	N	64	GP1	77	–	Vezza et al. 1977
					38	GP2	12	–	
Tacaribe	–	–	68	N	42	GP	79	–	Gard et al. 1977
Tamiami	–	–	66	N	44	GP	77	P	
Junin	–	–	60	N	44		–	–	Grau et al. 1981
					35–39				
Junin	–	–	54	VP-3(N)	91	VP-1	25	VP-6	Martinez, Segovia, and de Mitri 1977
					72	VP-2			
					55	VP-4			
					38	VP-5			
Machupo	–	–	68	N	50	GP1	84	–	Gangemi et al. 1978
					41	GP2	74	–	
							15		
Mopeia	180	L	61	N	48	GP1	84	–	Gonzalez et al. 1984
					35	GP2			
Mobala	180	L	60	N	48	GP1			
					37	GP2			
Lassa	180	L	61	N	45	GP1			
					38	GP2			
Lassa	–		60	N	45	GP1	76	–	Clegg and Lloyd 1983
					38	GP2	68	–	
							25[1]		
							20[1]		

[a] Nucleoprotein-related degradation or cleavage fragments

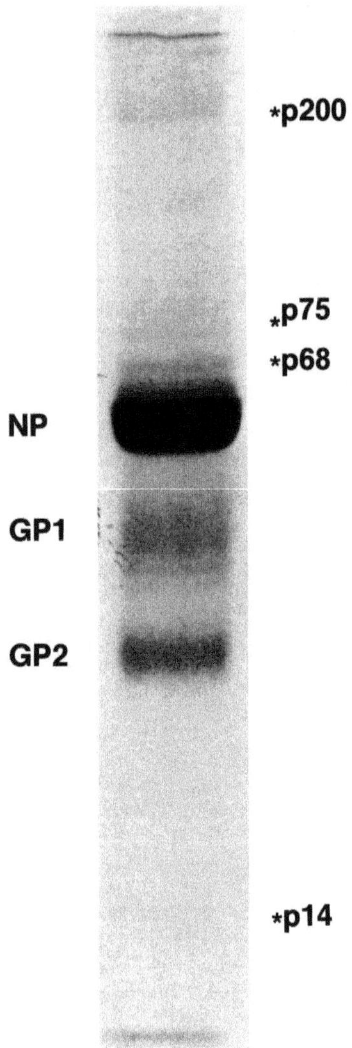

Fig. 1. [^{35}S]methionine-labeled polypeptides of LCMV. BHK-21 cells were infected with LCMV ARM at a multiplicity of infection of 0.1 PFU/cell then incubated for 24 h before adding 100 µCi/ml of [^{35}S]methionine in methionine-free medium. Supernatant medium was harvested at 40 h and virus purified and analyzed by SDS-PAGE as described (BUCHMEIER et al. 1978). Evident in the autoradiogram are the major viral polypeptides NP, GP1, and GP2, as well as quantitatively minor polypeptides (p) of 200 K, 75 K, 68 K, and 14 K. In other experiments we have shown that p 200 is encoded on the viral L RNA segment and that p 68 apparently shares tryptic peptides with NP. No definitive information is available for p 75 or p 14

1977), and several low molecular weight degradation products derived from nucleocapsid protein (HARNISH et al. 1981; BRUNS et al. 1983b; BUCHMEIER 1984). As detailed in BISHOP and AUPERIN, and SOUTHERN and BISHOP, this volume, molecular cloning approaches have definitively assigned NP and the GPC precursor of GP1 and GP2 to the S RNA segments of LCM and Pichinde viruses. These two open reading frames leave no room on S for additional primary translation products. Moreover, HARNISH et al. (1983) have mapped the 200 K L of Pichinde virus to the viral L RNA segment using genetic reassortants, and we (BUCHMEIER, SINGH, and SOUTHERN, unpublished observation) have shown directly, using synthetic peptide antibodies, that the L RNA segment of LCMV encodes a unique ca. 180 K L. Hence it appears clear that there is not sufficient coding capacity within the genome to accommodate all of the reported additional polypeptides and glycopeptides. Thus it is more likely, for example, that the additional glycoprotein components reported for LCMV by BRUNS et al. (1983b), which include glycoproteins of 130 K, 85 K, and 60 K, in addition to the previously documented 44 K (GP1) and 35 K (GP2) species, represent products of either atypical or incomplete posttranslational processing or contaminating cellular proteins. These differences will ultimately be resolved using well-defined monoclonal and sequence-specific, antipeptide antibodies and correlating the results with cDNA sequence analysis.

Localization of the structural proteins in the virion has been studies in detail. As with other enveloped RNA viruses, the glycopeptides are displayed on the external surface of the viral envelope. VEZZA et al. (1977) demonstrated that digestion of Pichinde virus with proteolytic enzymes produced "bald" or spikeless virions which were devoid of GP1 and GP2. GARD et al. (1977) confirmed this result for the glycoproteins of Tacaribe and Tamiami viruses, and BUCHMEIER et al. (1978) also showed that the GP1 and GP2 of LCMV were susceptible to proteolysis. Moreover by surface iodination of LCMV (BUCHMEIER and OLDSTONE 1979) it was shown that GP1 was the predominant virus-specific molecule accessible to lactoperoxidase-catalyzed iodination on the plasma membranes of infected cells and on the envelope. The macromolecular arrangement of glycoproteins in the membrane has been studied by BRUNS and LEHMANN-GRUBE (1983). Based primarily on cross-linking and nearest neighbor analyses, they suggest that the glycoprotein spike of LCMV consists of a complex of 1 gp 35 molecule linked to either three gp 44 molecules or one gp 44 and one gp 85 molecule. This model awaits confirmation, however, and must be viewed with caution since it fails to take into account the essentially equimolar concentrations of GP1 and GP2 (i.e., gp 44 and gp 35) in the membrane (VEZZA et al. 1977) and also incorporates two glycoproteins, gp 85 and gp 60, which have not been confirmed by others.

The nucleoproteins of arenaviruses reside predominantly in the ribonucleoprotein (RNP) core of the virion. The basic configuration of the Pichinde virus RNP was found by YOUNG and HOWARD (1983) to be a linear array of globular subunits, or nucleosomes, 4–5 nm in diameter, that represent individual molecules of the NP. This "beaded" filament appears to fold progressively through a number of 12–15 nm helical structures and these strands in turn were twisted to form 20 nm thick fibers seen in isolated core structures.

In addition to NP, we have also found L in association with the RNP. Preparations of purified LCM virions were subjected to disruption with 1% NP-40 and 0.5 M KCl, then pelleted through a sucrose layer to exclude soluble components. L was detected in the pelleted RNP complex by Western blotting with antipeptide antibodies to an L-RNA specific amino acid sequence.

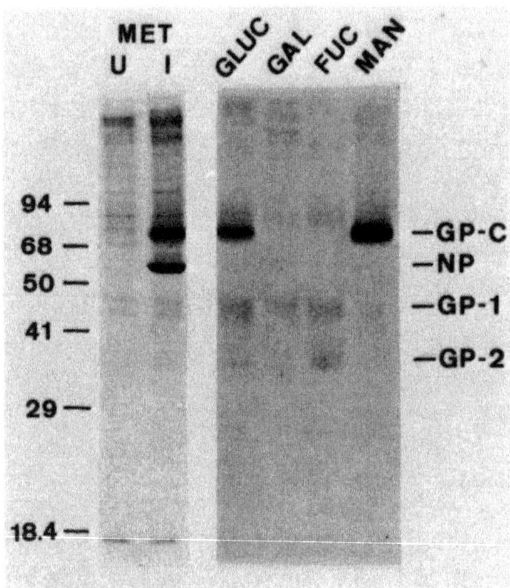

Fig. 2. Labeling of cell-associated proteins with [^{35}S]methionine (*MET*) and [^{3}H]sugars (*GLUC*, glucosamine; *GAL*, galactose; *FUC*, fucose; *MAN*, mannose). Cells were infected at a moi of 0.1 then pulse labeled for 2 h from 42–44 h after infection when virus production was maximal. Cytosols were prepared and immunoprecipitated with a hyperimmune guinea pig antiserum as described (BUCHMEIER and OLDSTONE 1979). Both GPC and NP are readily detectable using [^{35}S]methionine label, but only the glycopeptides label with sugar precursors. GPC is heavily labeled with mannose and glucosamine but not with galactose and fucose, demonstrating that it is of the high mannose type. Mannose residues are trimmed prior to cleavage of GPC into GP1 and GP2

A single nonstructural glycopeptide termed GPC has been described in cells infected with LCM (BUCHMEIER and OLDSTONE (1979), Pichinde (HARNISH et al. 1981), Tacaribe (SALEH et al. 1977), Lassa fever (CLEGG and LLOYD 1983), and Machupo (LUKASHEVICH and LEMESHKO 1985) viruses. As discussed in more detail in Sect. 4, GPC is the precursor of the structural glycoproteins. Intracellular GPC is an oligomannosyl rich glycopeptide which is processed posttranslationally by carbohydrate trimming and proteolytic cleavage prior to virus release (Fig. 2). Pulse-chase experiments (HARNISH et al. 1981; DIMOCK et al. 1982) have shown that the synthesis of GPC correlates closely with production of infectious virions. As GPC synthesis and consequent expression of viral glycoprotein at the plasma membrane increase, so does virus production. In turn, late in the infection when virus production wanes, GPC synthesis and surface expression of the structural glycoproteins diminish. The cellular site of GPC cleavage has not been precisely defined. In our work (BUCHMEIER and OLDSTONE 1979) we found only the mature GP1 and GP2 on the surface of infected cells and virions and suggested that cleavage occurs intracellularly; however, VAN DER ZEIJST et al. (1983a) have shown uncleaved precursor at the surface of acutely infected cells using a sensitive in situ surface iodination method. It is possible that this apparent discrepancy is a quantitative effect since we have recently observed a small quantity of uncleaved GPC in mature virions by Western blotting (PAREKH and BUCHMEIER, unpublished observation).

Table 2. Enzymatic activities associated with arenaviruses

Activity	Virus	Associated with	Reference
RNA polymerase	Pichinde	Viral RNP complex	CARTER et al. 1974; LEUNG et al. 1979
Poly-A polymerase	Pichinde	Ribosomes	LEUNG et al. 1979
Poly-U polymerase	Pichinde	Ribosomes	LEUNG et al. 1979
Protein kinase	LCM	Viral RNP complex	HOWARD and BUCHMEIER 1983; BRUNS et al. 1986

3 Enzymatic Activities Associated with Arenaviruses

Several enzymatic activities have been detected in association with purified arenaviruses (Table 2). In at least one of these, a viral RNA polymerase activity is thought to be obligatory for viral replication (CARTER et al. 1974; LEUNG et al. 1979). Polymerase activity has been studied most extensively in association with Pichinde virus, where it was shown to catalyze RNA synthesis from the viral genomic RNA template. The RNA polymerase was shown to be associated with the viral nucleocapsid complex, and was distinguishable on the basis of localization, as well as divalent cation requirements, from ribosome-associated, Mn^{2+}-dependent, poly-A polymerase and Mg^{2+}-dependent poly-U polymerase activities. The most likely candidate for the viral RNA polymerase is the nucleocapsid-associated L protein (180–200 K) which we have recently shown to be encoded by the L RNA. Characterization of the mode of transcription by this enzyme is hampered by its low activity and lability (LEUNG et al. 1979; BUCHMEIER et al. 1980b).

In addition to the RNA polymerase which appears to be virally encoded and the poly-A and -U polymerases which are ribosome associated and thus probably of host origin, HOWARD and BUCHMEIER (1983) have described a virion-associated protein kinase activity. This kinase preferentially phosphorylates serine and threonine residues in NP in the presence of a phosphate donor such as ATP (Fig. 3). The activity is internal, as indicated by its release from detergent solubilized virions, and is not stimulated by cyclic nucleotides. Kinase activity was shown by density gradient studies to be associated with the nucleocapsid core of the virion. Although it is not clear whether the kinase is a bonafide viral gene product or an encapsidated cellular constituent, it is conceivable that phosphorylation may play a role in the regulation of viral RNA polymerase activity or in virion assembly. Most attempts to demonstrate phosphorylated proteins in arenaviruses have been unsuccessful (review in COMPANS and BISHOP 1985). BRUNS et al. (1986) however, report that a fraction of the 63 K nucleoprotein is present in a soluble phosphorylated form termed p 63 E. The significance of this observation remains to be determined; these authors have suggested that p 63 E plays a regulatory role in the viral replicative process. Clearly the roles of the RNA polymerase (transcriptase) and protein kinase activities in

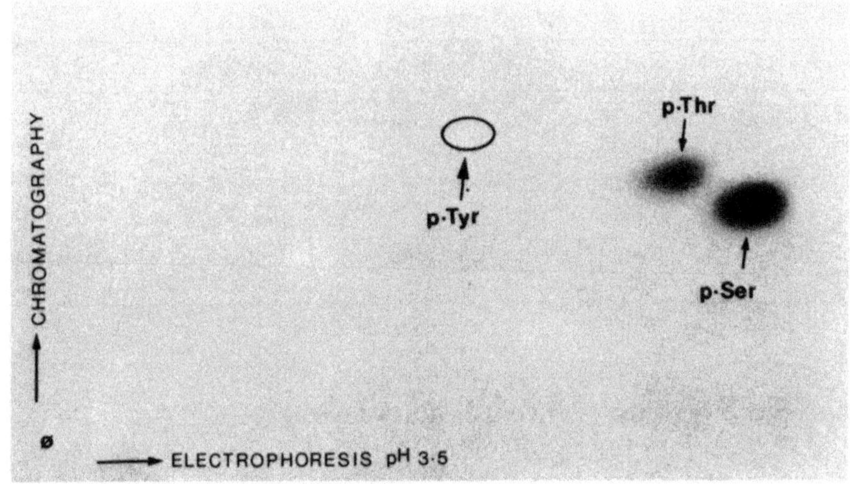

Fig. 3. Amino acids phosphorylated by the endogenous LCMV protein kinase. Phosphorylated NP was prepared as described (HOWARD and BUCHMEIER 1983) and hydrolyzed in 6N HCl. Samples were electrophoresed in the first dimension then chromatographed in the second. Radioactive residues were identified by comparison with unlabeled markers. Only phosphoserine (*p. Ser*) and phosphothreonine (*p.Thr*) were detected as phosphorylated products of the LCMV kinase. The position of phosphotyrosine (*p. Tyr*) is indicated for reference. Reproduced from HOWARD and BUCHMEIER (1983)

the replicative cycle of the virus, as well as definitive characterization of their molecular nature, remain subjects for future investigation.

4 Synthesis and Expression of Proteins

Arenaviruses have in common the capacity to establish persistent infections both in vivo and in vitro. Recent studies have suggested that the inherent ability of these viruses to regulate differentially the expression of NP and surface glycoproteins may play a role in the ability of the virus to evade immune surveillance in such persistent infections (WELSH and OLDSTONE 1977; WELSH and BUCHMEIER 1979; OLDSTONE and BUCHMEIER 1982; RODRIGUEZ et al. 1983). In general, information concerning the synthesis of the viral polypeptides is limited. In cells infected with LCMV, synthesis of NP, shown by radiolabeling with [^{35}S]methionine, is first detected 6 h after infection (BUCHMEIER et al. 1978), coincident with the start of the exponential phase of the replicative cycle. NP is apparently synthesized as a primary translation product since the native 63 K form is observed first in the infected cell. Pulse-chase experiments have not revealed higher molecular weight forms of it which might be candidates for a precursor molecule, although recent studies in a coupled in vitro transcription translation system have shown a 73 K translation product of the Tacaribe virus S RNA, which is apparently posttranslationally modified to yield the 68 K virion form of the molecule (BOERSMA and COMPANS 1985). Both the 68 K and 73 K polypeptides contained similar tryptic peptides, suggesting a common origin. We have observed a similar polypeptide of 68 K (Fig. 1) in LCMV and have shown in unpublished work that this molecule shares tryptic peptides with NP; caution must be exercised, however, since the 68 K protein may be contaminated with significant amounts of the more abundant 63 K NP. Nevertheless the possibility exists that a portion of NP message is translated as an elongated "read through" polypeptide or alternatively that NP (63 K) is rapidly cleaved posttranslationally off a larger primary translation product.

Several groups have shown that NP is present in both full length and degraded form in virions and infected cells. Pichinde virus was shown to encode polypeptides of 15 K and 20 K in infected cells, and these were antigenically related to NP (Buchmeier et al. 1977). Harnish et al. (1981) described NP-related polypeptides of 48 K, 38 K, and 28 K in Pichinde-infected BHK-21 cells, and, after a 3-h chase period, additional species of 17 K, 16.5 K, and 14 K were also evident. All of these six quantitatively minor polypeptides shared common tryptic peptides with NP suggesting their derivation by posttranslational proteolytic cleavage, although the potential presence of premature termination products or of mutants containing deletions in the NP gene was not formally precluded.

Young et al. (1985b) have observed similar NP-related cleavage products appearing late in Pichinde virus infection of Vero cells and have suggested that the intracellular level of NP plays a role in regulating genome replication transcription by committing newly transformed RNA to either nucleocapsid assembly or further rounds of replication. It was proposed that differences in cleavage patterns were related to the extent of virus regulation observed. At least one of the NP-related cleavage products, a 28 K fragment identifiable with a monoclonal antibody, was found in the cell nucleus late in the infection (Young et al. 1985a). These studies suggest the possibility that NP or cleavage fragments of it may serve as regulatory molecules to modulate transcription and/or replication.

Observations of cleavage fragments of nucleocapsid protein are not restricted to Pichinde virus. Clegg and Lloyd (1983), working with Lassa virus, reported N-related polypeptides of 36 K and 24 K, termed fN1 and fN2, respectively. In this instance fN1 and fN2 were thought to be artifacts resulting from proteolysis upon disruption of Lassa-infected cells; only full length N was detected by Western blotting in cells lysed under stringent denaturing conditions. Bruns et al. (1983b) have reported 25 K and 38 K polypeptides in LCM virions, which appeared in one-dimensional tryptic maps to be related to the NP (termed p 63 in their nomenclature). We have shown, using both peptide mapping and monoclonal antibody analyses, that LCMV contains variable amounts of NP-related polypeptides of 25 K, 38 K and approximately 40 K (Buchmeier, unpublished observations). The appearance and quantities of these species are highly variable and affected by factors such as the host cell used to cultivate the virus and the time after infection; hence, it is not clear at the moment whether these molecules play a functional role in virus replication.

A clearer picture of the synthesis and potential mechanisms of regulation of arenavirus glycoproteins has emerged from recent studies of their genetic

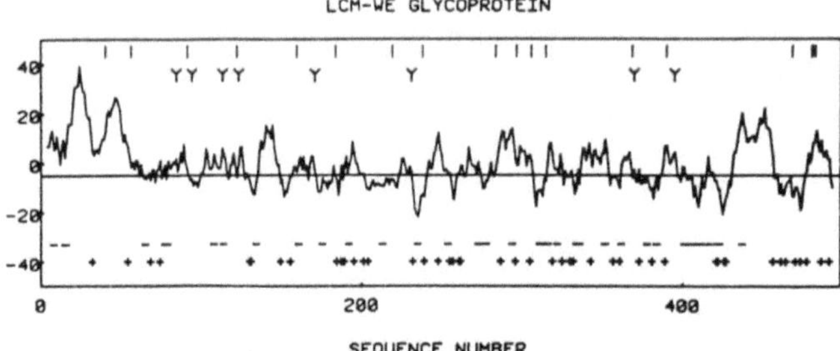

Fig. 4. Hydropathic plot, cysteine, and charged amino acid distribution for the GPC predicted gene product of LCMV-WE S RNA. Regions of the predicted protein with net hydrophobicity (areas *above* the center line) and those with net hydrophilicity (*below* the line) are displayed according to KYTE and DOOLITTLE (1982). As well as the positions of charged amino acids, cysteine residues (*vertical bars*) and potential asparagine-linked glycosylation sites (*Y*) are shown. Reproduced from ROMANOWSKY and BISHOP (1985)

structure (AUPERIN et al. 1984; ROMANOWSKI and BISHOP 1985; SOUTHERN et al. 1986). Previous investigation at the biochemical level demonstrated the presence of a high molecular weight (70–78 K) precursor of the virus structural glycoproteins in cells infected with LCM (BUCHMEIER et al. 1978; BUCHMEIER and OLD-STONE 1979), Tacaribe (SALEH et al. 1979), Pichinde (HARNISH et al. 1981), Lassa fever (CLEGG and LLOYD 1983), and Machupo (LUKASHEVICH and LEMESHKO 1985) viruses. Availability of nucleotide and amino acid sequence information has allowed mapping of the precursor glycoprotein, termed GPC for LCMV (BUCHMEIER et al. 1978), to the 5′ half of the virion S RNA strand (see BISHOP and AUPERIN, and SOUTHERN and BISHOP, this volume). Recent experiments in our laboratory have been directed at determining the fine structure of the GPC gene, orientation of the structural glycoproteins on the precursor, signals for proteolytic cleavage, and defining important antigenic regions of the molecule. Hydrophobicity profiles have been determined by the method of KYTE and DOOLITTLE (1982) for Pichinde virus and LCMV (Fig. 4). The GPC precursors of both LCMV WE and ARM consist of 498 amino acids in a single open reading frame. From the hydrophobicity profile it is evident that the protein contains three significant hydrophobic domains approximately spanning residues 1–32, 34–55, and 433–460 from the amino terminus. The first of these is throught by analogy with other membrane glycoproteins to constitute a signal sequence, while the carboxyl domain (433–460) is likely to constitute a membrane anchor. Amino acid sequence comparisons with the corresponding GPC genes of Pichinde and Lassa fever viruses have revealed that the gene consists of two domains. The amino terminal half is potentially highly glycosylated with six asn-x-ser/thr glycosylation sites in LCMV ARM and WE, and eleven sites in the corresponding region of Pichinde GPC. Work in our laboratory has shown that at least four of these six potential sites in LCMV are actually glycosylated. Further, this domain shows only a low degree of amino acid homology between LCMV and Pichinde virus. In contrast, the carboxyl domain contains sequences which contain fewer glycosylation sites (three in LCM, five

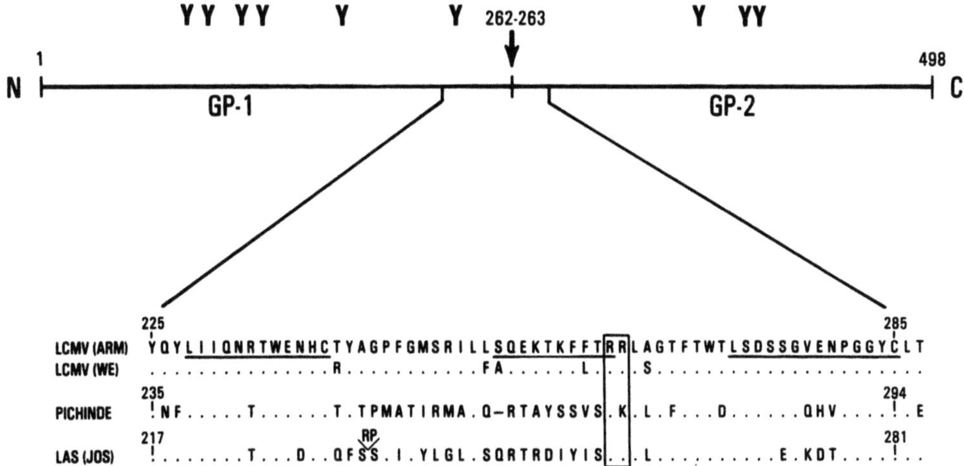

Fig. 5. Cleavage region of the GPC precursors of LCMV, Pichinde, and Lassa fever viruses. The linear sequence of LCMV-ARM GPC is represented with glycosylation sites marked (*Y*). Amino acids flanking the putative cleavage site in LCMV-ARM, LCMV-WE, Pichinde (AUPERIN et al. 1984), and Lassa fever (Josiah) (D. AUPERIN, personal communication) viruses are shown with alignment for maximum homology. Conserved amino acids relative to LCMV-ARM are indicated by *dots,* and peptides used to map the cleavage site as described in the text are *marked.* Note that the position of the conserved pair of basic amino acids relative to the *underlined* conserved flanking sequences is retained in all four viruses

in Pichinde) and share a high degree of homology with the corresponding sequences in Pichinde virus. We have used synthetic peptide and monoclonal antibodies to study the coding assignments and topography of the LCMV GPC gene. Peptides were synthesized corresponding to four regions of GPC (amino acids 59–79, 228–239, 272–285, and 378–391), and site-specific antisera raised in rabbits. Antibodies made against the more amino terminal peptides 59–79 and 228–239 reacted with native and denatured GP1, while antisera to the two peptides closest to the carboxyl end (271–285 and 378–391) reacted with GP2. Thus the gene order on the GPC message is NH_2-(GP1) (GP2)–COOH. Examination of the sequence between peptides 228–239 and 272–285 revealed the presence of a double basic amino acid sequence Arg-Arg at residues 262–263 flanked by hydrophobic amino acids. Alignment with the sequences of Pichinde (AUPERIN et al. 1984) and Lassa GPC (AUPERIN, personal communication) showed corresponding paired basic sequences in both of those viruses (Fig. 5). On the basis of these data it is apparent the cleavage site utilized on GPC to yield the GP1 and GP2 is defined by this conserved double basic amino acid sequence Arg-Arg on LCM and Lassa fever viruses, and Arg-Lys on Pichinde virus. Thus this site is likely to be a general feature of the arenavirus group. Moreover, analysis of sequence homology among the GPC precursors of these viruses has shown extensive conservation of sequences in GP2 predicted by the earlier demonstration of conserved antigens in GP2 (BUCHMEIER et al. 1980a, 1981, 1984) using monoclonal antibodies.

5 Antigenic Topography of Arenavirus Glycoproteins

Arenaviruses differ in their susceptibility to antibody-mediated neutralization. Neutralizing antibodies to LCMV have been shown to be directed against the GP1 (gp 44) glycoprotein (BUCHMEIER et al. 1981; BUCHMEIER 1984; BRUNS et al. 1983a; PAREKH and BUCHMEIER 1986). Similarly, monoclonal antibodies against the single glycoprotein of Tacaribe virus-mediated, highly efficient virus neutralization (ALLISON et al. 1984; HOWARD et al. 1985). Moreover, using competitive binding assays and analysis of neutralization resistant mutants, it was possible to map two distinct epitopes on Tacaribe G (HOWARD et al. 1985). One epitope, characterized by four monoclonal antibodies, was the target of highly efficient neutralization, while a single antibody to a second site was less efficient, leaving a large non-neutralizable persistent fraction. Failure to neutralize was not likely to be due to virus aggregation since addition of a second antibody to the alternate site resulted in further reduction in virus titer. Analysis of neutralization kinetics for the highly efficient monoclonal antibody suggested that the reaction followed double hit kinetics.

We have assessed the antigenic topography of the LCMV glycoproteins using a large library of monoclonal antibodies against GP1 and GP2 to map the epitopes on these molecules (PAREKH and BUCHMEIER 1986). Elicitation of neutralizing monoclonal antibodies to LCMV in the BALB/c mouse was a relatively infrequent event. Only 6 of 46 antibodies to the LCMV glycoproteins neutralized virus infectivity in vitro. Five of these antibodies were raised against the WE strain of virus and mapped by competition binding assay to a single conformation-dependent epitope (GP1A) shared by both ARM and WE as well as other LCMV strains (Fig. 6). The sixth neutralizing MAb was uniquely specific for the LCMV-ARM strain and its binding to that strain was only marginally affected by the other five antibodies, suggesting binding to a topographically related but not identical epitope (GP1D). Nonneutralizing MAb were found to be directed against two additional sites on GP1, (GP1B, C) as well as against three sites on GP2. The relevance of these data to the polyclonal antibody response was investigated using a potent neutralizing antiserum raised in guinea pigs. This antibody reacted predominantly with conformation-dependent structures on GP1, as indicated by its failure to bind in Western blotting, and its binding was completely inhibited by any of the five LCMV WE-specific neutralizing MAb against site GP1A. These results imply that the LCMV WE GP1 has a single immunodominant neutralizing antigenic determinant (GP1A) and that the LCMV ARM strain bears an additional topographically related but not identical site (GP1D). Attempts to neutralize other arenaviruses have met with mixed success. Patient and experimentally produced antisera show potent neutralizing activity against Junin virus; however, similar reagents from Lassa fever convalescent patients show rather low neutralizing potency unless complement is added to potentiate the effects of antibody (PETERS 1984). Virus neutralization is discussed in depth in one chapter by C.R. HOWARD in the accompanying volume in this series (134). From the brief treatment here it is evident that more information about the molecular nature of neutralizing antigenic determinants of arenaviruses is necessary before rational approaches to immunotherapy and immunization can be made. Obviously one needs to

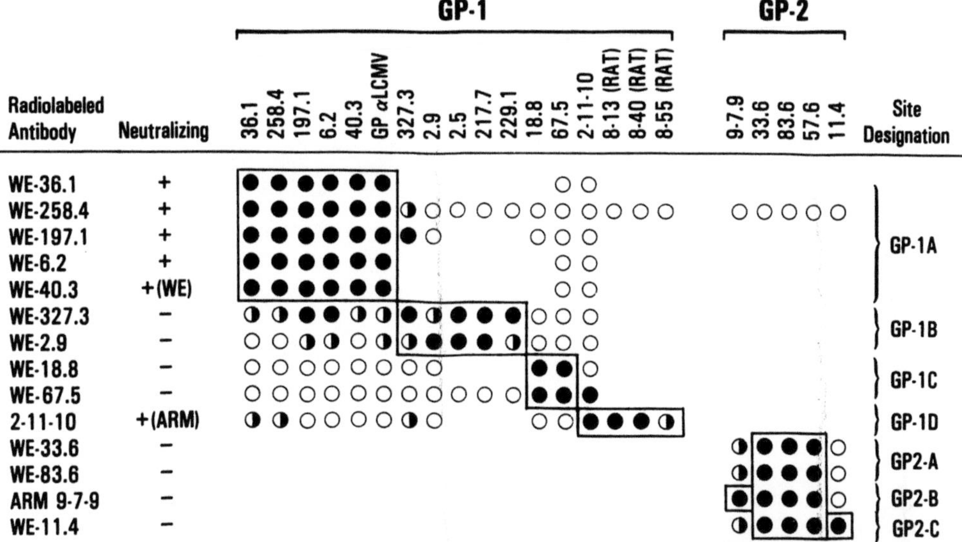

Fig. 6. Summary of results of competitive binding assay to map the epitopes on LCMV glycoproteins. Dilutions $(10^{-1}-10^{-7})$ of unlabeled competing antibody were tested for their capacity to inhibit binding of constant amounts of purified, radiolabeled monoclonal antibodies. Combinations yielding greater than 80% inhibition are indicated by *filled circles*, 40%–80% by *half-filled circles* and less than 40% by *open circles*. Using this assay four epitopes on GP1 of LCMV-ARM (three on GP1 of LCM-WE) and three on GP2 have been defined. One of these, GP1A, was the major virus-neutralizing site recognized by mouse monoclonal and guinea pig polyclonal (GP anti-LCM) antibodies to LCMV. Note also that the LCMV-ARM-specific site GP1D was recognized by neutralizing monoclonal antibodies produced against that virus in the rat (8–13, 8–40, and 8–55). On GP2 antibodies WE 33.6 and 83.6 recognized a site termed GP2A which is common to both Old and New World arenaviruses (BUCHMEIER 1984)

define structure which will elicit strong protective immune responses without the risk of triggering immunopathologic disease.

6 Pathobiological Role of Specific Viral Gene Products In Vivo

Viral polypeptides and their degradation products trigger many of the pathobiologic manifestations observed in arenavirus infections. In the lifelong persistent infection of mice with LCMV a wasting syndrome has been well documented which is characterized by the development of immune complexes composed of viral antigen and antiviral antibody (OLDSTONE et al. 1980, 1983). These complexes lodge in the renal glomeruli where they trigger a chronic glomerulonephritis. At least one component of the virus has been identified in the glomeruli of diseased mice. Using a monospecific antibody to the NP of LCMV, BUCHMEIER and OLDSTONE (1978) demonstrated colocalization of NP antigen and the host glomerular mesangium.

A role of NP in neuronal dysfunction has also been proposed to occur during LCMV persistence. RODRIGUEZ et al. (1983) observed expression of NP in association with polyribosomes in the cytoplasm of neurons from widespread

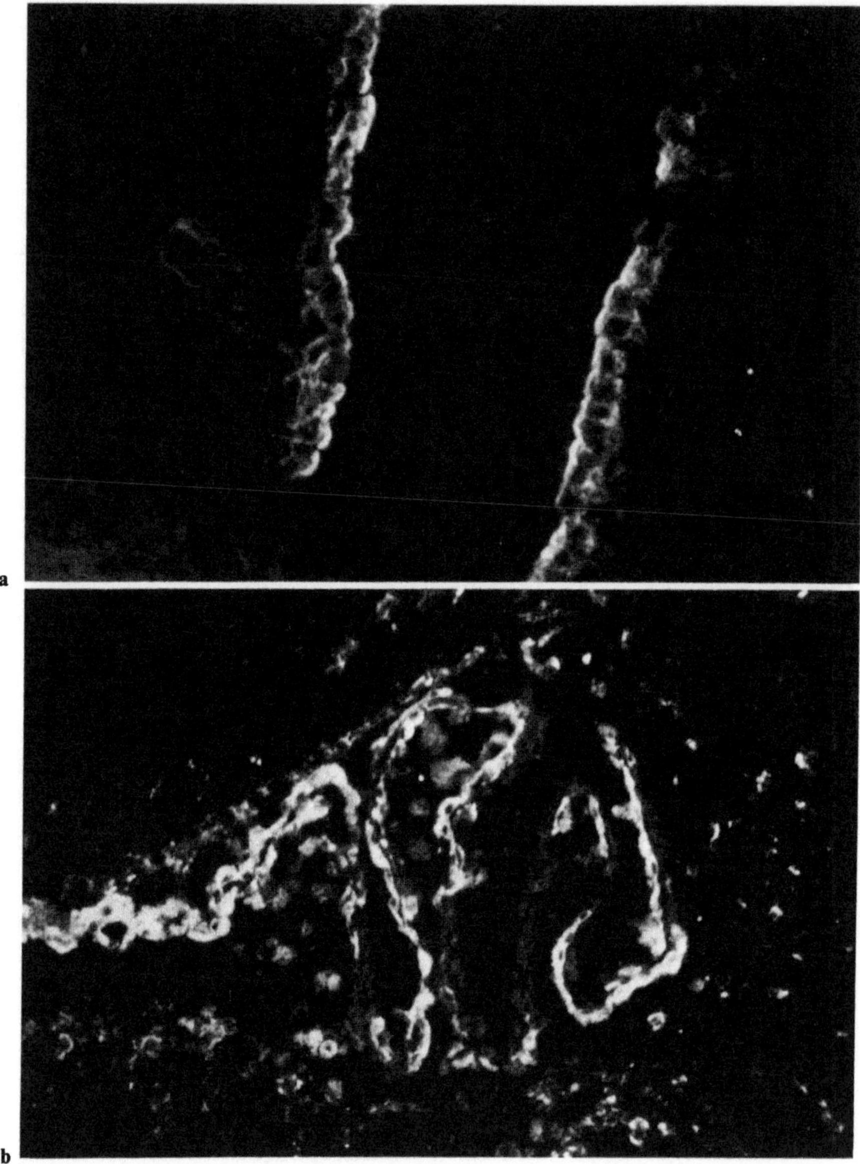

Fig. 7a, b. Immunofluorescent staining of LCMV GP1 (a) on ependymal cells lining a ventricule and (b) the choroid plexus of a C57B1/6 mouse infected 6 days earlier with LCMV-ARM. Staining with FITC-conjugated monoclonal antibody to GP1. Original magnification × 250

areas of the CNS. In contrast, these workers found no significant expression of viral glycoproteins. It was speculated that the presence of NP on the neuronal polyribosomes compromised their function (see FRANCIS et al., this volume).

Acute LCMV infection following intracerebral inoculation results in a fatal choriomeningitis (reviewed in BUCHMEIER et al. 1980b). We have used monoclonal antibodies against individual virus structural proteins to study their expression in the CNS following acute infection (BUCHMEIER and KNOBLER 1984). As is evident from Fig. 7, viral GP1 is expressed on the apical surfaces of ependymal cells in the CNS. At this site the glycoprotein (and perhaps also other virally coded proteins) triggers the well-characterized immune response which results in choriomeningitis and death.

Finally, the role of the viral L gene-encoded proteins in pathogenesis has recently been explored by RIVIERE et al. (1985, this volume) using genetic reassortants between strains of LCMV that differed in virulence for guinea pigs. These workers demonstrated that L RNA-encoded products were necessary for expression of the pathogenic potential of the virus. Each of these topics will be examined in detail in later chapters. It is clear from such studies that understanding the molecular basis of viral persistence and pathogenesis of arenaviruses is an attainable goal.

Acknowledgements. This is publication number 4477-IMM from the Department of Immunology, Scripps Clinic and Research Foundation. This study was supported by USPHS grant No. AI 16102 and training grant No. AG00080. We thank DONNA URANOWSKI for manuscript preparation.

References

Allison LM, Salter M, Buchmeier MJ, Lewicki H, Howard CR (1984) Neutralization of arenaviruses: reaction of Tacaribe virus and variants with monoclonal antibodies. In: Compans RW, Bishop DHL (eds) Segmented negative strand viruses, arenaviruses, bunyaviruses, and orthomyxoviruses. Academic, New York, pp 209–216

Auperin DD, Romanowski V, Galinski M, Bishop DHL (1984) Sequencing studies of Pichinde arenavirus S RNA indicate a novel coding strategy, an ambisense viral S RNA. J Virol 52 (3):897–904

Blount P, Elder J, Lipkin WI, Southern PJ, Buchmeier MJ, Oldstone MBA (1986) Dissecting the molecular anatomy of the nervous system: analysis of RNA and protein expression in whole body sections of laboratory animals. Brain Res 382:257–265

Boersma DP, Compans RW (1985) Synthesis of Tacaribe virus polypeptides in an in vitro coupled transcription and translation system. Virus Res 2:261–271

Bro-Jorgensen K (1971) Characterization of virus-specific antigen in cell culture infected with lymphocytic choriomeningitis virus. Acta Pathol Microbiol Scand [B] 79:466–474

Brown WJ, Kirk BE (1969) Complement-fixing antigen from BHK-21 cell cultures infected with lymphocytic choriomeningitis virus. Appl Microbiol 18:496–499

Bruns M, Lehmann-Grube F (1983) Lymphocytic choriomeningitis virus: V. Proposed structural arrangement of proteins in the virion. J Gen Virol 64:2157–2167

Bruns M, Cihak J, Muller G, Lehmann-Grube F (1983a) Lymphocytic choriomeningitis virus: VI. Isolation of a glycoprotein mediating neutralization. Virology 130:247–251

Bruns M, Peralta LM, Lehmann-Grube F (1983b) Lymphocytic choriomeningitis virus: III. Structural proteins of the virion. J Gen Virol 64:599–611

Bruns M, Zeller W, Rohdewohld H, Lehmann-Grube F (1986) Lymphocytic choriomeningitis virus: IX. Properties of the nucleocapsid. Virology 151:77–85

Buchmeier MJ (1984) Antigenic and structural studies on the glycoproteins of lymphocytic choriomeningitis virus. In: Compans RW, Bishop DHL (eds) Segmented negative strand viruses, arenaviruses, bunyaviruses, and orthomyxoviruses. Academic, New York, pp 193–200

Buchmeier MJ, Knobler RL (1984) Experimental models for immune-mediated and immune-modulated diseases. In: Behan P, Spreafico F (eds) Neuroimmunology. Raven, New York, pp 219–227

Buchmeier MJ, Oldstone MBA (1977) Identity of the viral protein responsible for serologic cross reactivity among the Tacaribe complex arenaviruses. In: Mahy BWJ, Barry RD (eds) Proc of Conf on Negative Strand Viruses and the Host Cell. Academic, London, pp 91–96

Buchmeier MJ, Oldstone MBA (1978) Virus-induced immune complex disease: identification of specific viral antigens and antibodies deposited in complexes during chronic lymphocytic choriomeningitis virus infection. J Immunol 120(4):1297–1304

Buchmeier MJ, Oldstone MBA (1979) Protein structure of lymphocytic choriomeningitis virus: evidence for a cell-associated precursor of the virion glycopeptides. Virology 99:111–120

Buchmeier MJ, Gee SR, Rawls WE (1977) Antigens of Pichinde virus: I. Relationship of soluble antigens derived from infected BHK-21 cells to the structural components of the virion. J Virol 22(1):175–186

Buchmeier MJ, Elder JH, Oldstone MBA (1978) Protein structure of lymphocytic choriomeningitis virus: identification of the virus structural and cell associated polypeptides. Virology 89:133–145

Buchmeier MJ, Lewicki H, Tomori O, Johnson KM (1980a) Monoclonal antibodies to lymphocytic choriomeningitis virus react with pathogenic arenaviruses. Nature 288:486–487

Buchmeier MJ, Welsh RM, Dutko FJ, Oldstone MBA (1980b) The virology and immunobiology of lymphocytic choriomeningitis virus infection. Adv Immunol 30:275–331

Buchmeier MJ, Lewicki HA, Tomori O, Oldstone MBA (1981) Monoclonal antibodies to lymphocytic choriomeningitis and Pichinde viruses: generation, characterization, and cross-reactivity with other arenaviruses. Virology 113:73–85

Carter MF, Biswal N, Rawls WE (1974) Polymerase activity of Pichinde virus. J Virol 13(3):577–583

Chastel C (1970) Immunodiffusion studies on a fluorocarbon-extracted antigen of lymphocytic choriomeningitis virus. Acta Virol (Praha) 14:507–509

Clegg JAC, Lloyd G (1983) Structural and cell associated proteins of Lassa virus. J Gen Virol 64:1127–1136

Compans RW, Bishop DHL (1985) Biochemistry of arenaviruses. Curr Top Microbiol Immunol 114:153–175

Dimock K, Harnish DG, Sisson G, Leung W-C, Rawls WE (1982) Synthesis of virus-specific polypeptides and genomic RNA during the replicative cycle of Pichinde virus. J Virol 43(1):272–283

Gard GP, Vezza AC, Bishop DHL, Compans RW (1977) Structural proteins of Tacaribe and Tamiami virions. Virology 83:84–95

Grau O, Franze-Fernandez M, Romanowski V, Rustici S, Roasas MF (1981) Junin Virus Structure. In: Bishop DHL, Compans RW (eds) The Replication of Negative Strand Viruses. Elsevier North Holland, New York, pp 11–14

Gschwender HH, Rutter G, Lehmann-Grube F (1976) Lymphocytic choriomeningitis virus II. Characterization of extractable complement-fixing activity. Med Microbiol Immunol (Berl) 162:119–131

Harnish DG, Leung WC, Rawls WE (1981) Characterization of polypeptides immunoprecipitable from Pichinde virus-infected BHK-21 cells. J Virol 38:840–848

Harnish DG, Dimock K, Bishop DHL, Rawls WE (1983) Gene mapping in Pichinde virus: assignment of viral polypeptides to genomic L and S RNAs. J Virol 46:638–641

Howard CR, Buchmeier MJ (1983) A protein kinase activity in lymphocytic choriomeningitis virus and identification of the phosphorylated product using monoclonal antibody. Virology 126:538–547

Howard CR, Lewicki H, Allison L, Salter M, Buchmeier MJ (1985) Properties and characterization of monoclonal antibodies to Tacaribe virus. J Gen Virol 66:1383–1395

Kyte J, Doolittle RF (1982) A simple method for displaying the hydropathic character of a protein. J Mol Biol 157:105–132

Lehmann-Grube F, Slencyka W, Tees R (1969) A persistent and inapparent infection of L cells with the virus of lymphocytic choriomeningitis. J Gen Virol 5:63–81

Leung W-C, Leung MFKL, Rawls WE (1979) Distinctive RNA transcriptase, polyadenylic acid polymerase, and polyuridylic acid polymerase activities associated with Pichinde virus. J Virol 301:98–107

Lukashevich IS, Lemeshko NN (1985) Machupo virus polypeptides: identification by immunoprecipitation. Arch Virol 86:85–99

Oldstone MBA, Buchmeier MJ (1982) Restricted expression of viral glycoprotein in cells of persistently infected mice. Nature 300:360–362

Oldstone MBA, Buchmeier MJ, Doyle MV, Tishon A (1980) Virus-induced immune complex disease: specific anti-viral antibody and Clq binding material in the circulation during persistent lymphocytic choriomeningitis virus infection. J Immunol 124(2):831–838

Oldstone MBA, Tishon A, Buchmeier MJ (1983) Virus-induced immune complex disease: genetic control of Clq binding complexes in the circulation of mice persistently infected with lymphocytic choriomeningitis virus. J Immunol 130(2):912–918

Parekh BS, Buchmeier MJ (1986) Proteins of lymphocytic choriomeningitis virus: antigenic topography of the viral glycoproteins. Virology 153:168–178

Pedersen IR (1973) LCM virus: its purification and its chemical and physical properties. In: Lehmann-Grupe F (ed) Lymphocytic choriomeningitis virus and other arenaviruses. Springer, Berlin Heidelberg New York, pp 13–23

Peralta L, Lehmann-Grube F (1983) Properties of lymphocytic choriomeningitis virus interfering particles. Arch Virol 77(1):61–69

Peters CJ (1984) Arenaviruses. In: Belshe RG (ed) Textbook of Human Virology. PSG Publishing, Littleton, MA

Ramos BA, Courtney RJ, Rawls WE (1972) Structural proteins of Pichinde virus. J Virol 10:661–667

Rawls WE, Buchmeier MJ (1975) Arenaviruses: purification and physicochemical nature. Bull WHO 52:393–401

Rawls WE, Leung W-C (1979) Arenaviruses. Compr Virol 14:157–192

Riviere Y, Ahmed R, Southern PJ, Buchmeier MJ, Oldstone MBA (1985) Genetic mapping of lymphocytic choriomeningitis virus pathogenicity: virulence in guinea pigs in associated with the L RNA segment. J Virol 55:704

Rodriguez M, Buchmeier MJ, Oldstone MBA, Lampert PW (1983) Ultrastructural localization of viral antigens in the CNS of mice persistently infected with lymphocytic choriomeningitis virus (LCMV). Am J Pathol 110:95–100

Romanowski V, Matsuura Y, Bishop DHL (1985) Complete sequence of the S RNA of lymphocytic choriomeningitis virus (WE strain) compared to that of Pichinde arenavirus. Virus Res 3:101–114

Saleh F, Gard GP, Compans RW (1979) Synthesis of Tacaribe viral proteins. Virology 93:369–376

Simon M (1970) Multiplication of lymphocytic choriomeningitis in various systems. Acta Virol (Praha) 14:369–376

Smadel JE, Baird RD, Wall MJ (1939) A soluble antigen of lymphocytic choriomeningitis: I. Separation of soluble antigen from virus. J Exp Med 70:53–66

Smadel JE, Wall MJ, Baird RD (1940) A soluble antigen of lymphocytic choriomeningitis. II. Characteristics of the antigen and its use in precipitin reactions. J Exp Med 71:43–53

Smadel JE, Green RH, Paltauf PM, Gonzales TA (1942) Lymphocytic choriomeningitis: two human fatalities following an unusual febrile illness. Proc Soc Exp Biol Med 49:683–686

Southern PJ, Buchmeier MJ, Ahmed R, Francis SJ, Parekh B, Riviere Y, Singh MK, Oldstone MBA (1986) Molecular pathogenesis of arenavirus infection. In: Lerner RA, Chanock RM, Brown F (eds) Vaccines 86: Modern Approaches to Vaccines. Cold Spring Harbor Laboratory, New York, pp 239–245

Traub E (1936) Persistence of lymphocytic choriomeningitis virus in immune animals and its relation to immunity. J Exp Med 63:847–861

van der Zeijst BAM, Bleumink N, Crawford LV, Swyryd EA, Stark GR (1983a) Viral proteins and RNAs in BHK cells persistently infected by lymphocytic choriomeningitis virus. J Virol 48:262–270

van der Zeijst BAM, Noyes BE, Mirault M-E, Parker B, Osterhaus ADME, Swyryd EA, Bleumink N, Horzinek MC, Stark GR (1983b) Persistent infection of some standard cell lines by lymphocytic choriomeningitis virus: transmission of infection by an intracellular agent. J Virol 48:249–261

Vezza AC, Gard GP, Compans RW, Bishop DHL (1977) Structural components of the arenavirus Pichinde. J Virol 23:776–786

Welsh RM, Buchmeier MJ (1979) Protein analysis of defective interfering lymphocytic choriomeningitis virus and persistently infected cells. Virology 96:503–515

Welsh RM, Oldstone MBA (1977) Inhibition of immunologic injury of cultured cells infected with lymphocytic choriomeningitis virus: role of defective interfering virus in regulating viral antigenic expression. J Exp Med 145:23–29

Wilsnack RE, Rowe WP (1964) Immunofluorescent studies of the histopathogenesis of lymphocytic choriomeningitis virus infection. J Exp Med 120:829–841

Wulff H, Lange JV, Webb PA (1978) Interrelationships among arenaviruses measured by indirect immunofluorescence. Intervirology 9:344–350

Young PR, Howard CR (1983) Fine structure analysis of Pichinde virus nucleocapsids. J Gen Virol 64:833–842

Young PR, Chanas AC, Howard CR (1985a) Localization of arenavirus proteins in the nuclei of infected cells. Virus Res (Suppl) 1:43

Young PR, Lee SR, Howard CR (1985b) Regulation of Pichinde virus replication in Vero and BHK-21 cells. Proc Fourth Inter Symposium of the Heinrich-Pette Institute for Experimentale Virology Hamburg, September

Mapping Arenavirus Genes Causing Virulence

Y. Rivière

1 Introduction

The reassortment of genetic information from RNA viruses with segmented genomes, i.e., Arenaviridae, Bunyaviridae, Orthomyxoviridae, and Reoviridae families, can occur within cells infected by two different viral strains from the same family (Compans et al. 1981; Fields and Greene 1982; Webster et al. 1982; Palese 1984). The phenomenon of reassortment also occurs in vivo and has been exploited to define the function of viral genes in pathogenesis (Fields and Greene 1982; Webster et al. 1982).

Extensive information is available today on the biological properties of arenaviruses (Lehmann-Grube 1971; Buchmeier et al. 1980), but 2 years ago very little was known about the function of their genes in pathogenicity (Compans and Bishop 1985). Lymphocytic choriomeningitis virus (LCMV) is the prototype arenavirus and contains two single strand RNA species, large (L) and small (S), with approximate molecular weights of 2.8×10^6 and 1.3×10^6, respectively (Pedersen 1971; Dutko and Oldstone 1983; Compans and Bishop 1985). The following four major viral structural proteins have been identified: an internal nucleoprotein, NP (63K), associated with the genomic RNA; two surface glycoproteins, GP1 (44K) and GP2 (35K) (Buchmeier and Oldstone 1981); and a large protein, L (180K) (Buchmeier et al. 1985). To define the precise role of gene products of the L and S RNA segments in disease, we have generated reassortant viruses between various strains of LCMV. Following coinfection of cells with two different LCMV strains, recombinants are generated by reassortment of genome segments. The Armstrong (ARM) and WE

Oncologie Virale, Institut Pasteur, 28, rue du Dr. Roux, 75015 Paris, France

Current Topics in Microbiology and Immunology, Vol. 133
© Springer-Verlag Berlin · Heidelberg 1987

strains of LCMV were chosen as the parental viruses for generating the reassortants because the NP and glycoproteins of the ARM and WE strains can be readily distinguished by monoclonal antibodies and tryptic peptide mapping (BUCHMEIER 1984). Using monoclonal antibodies specific for the NP and glycoproteins of WE and ARM, we have shown that the S RNA segment of LCMV codes for the three major structural polypeptides GP1, GP2, and NP (RIVIERE et al. 1985a). Similar results have been reported for Pichinde virus (VEZZA et al. 1980; COMPANS et al. 1981; LEUNG et al. 1981; HARNISH et al. 1983), another member of the arenavirus family, suggesting that different arenaviruses may have similar gene coding strategies.

2 Use of Reassortant Viruses to Map the RNA Segment(s) Involved in Lymphocytic Choriomeningitis

Several individual LCMV strains have been distinguished and segregated on the basis of three types of analysis: hybridization with cDNA probes, T1 oligonucleotide mapping, and monoclonal antibody analysis (DUTKO and OLDSTONE 1983; BUCHMEIER 1984; SOUTHERN et al. 1984). In addition, variations in the biological properties of different LCMV strains have been observed. We will describe briefly three of these experimental models. C3H newborn mice inoculated with the ARM CA1371 strain of LCMV become retarded in growth and low in blood sugar due to viral curtailment of growth hormone synthesis, but infection with the WE strain has no effect on growth, development, or blood sugar level (OLDSTONE et al. 1985). Adult guinea pigs resist infection by the ARM strain, but succumb to infection by the WE strain (LEHMANN-GRUBE 1971). LCMV-specific cytotoxic T-lymphocytes (CTL) can selectively distinguish between strains of LCMV when studied in mice of the H-2^d haplotype (AHMED et al. 1984).

2.1 Mapping of the Growth Hormone Defect to the S RNA of the LCMV Genome

Recent experiments performed by Oldstone have shown that LCMV can infect growth hormone-producing cells in the anterior lobe of the pituitary gland (OLDSTONE et al. 1982). This infection, associated with a reduced level of growth hormone, growth retardation, and hypoglycemia, frequently leads to death. However, despite the viral replication and the perturbation of growth hormone production, the infected cells remain free from structural injury. Several murine strains infected at birth with any of various LCMV strains persistently carry virus throughout their lives. Growth hormone insufficiency associated with glucose imbalance develops only when a susceptible genetic background is combined with infection by a specific LCMV strain. This is the case with the C3H/St mice infected at birth with LCMV ARM CA 1371. These mice exhibit marked growth retardation and hypoglycemia, and they rarely survive beyond their thirtieth day of life. The amount of growth hormone synthesized in the pituitaries of these infected mice is markedly lower than in uninfected controls, and over 85% of the growth hormone-containing cells from infected mice replicate LCMV. In contrast, C3H/St mice neonatally infected with LCMV WE strain or BALB/c WEHI and SWR/J mice similarly infected with LCMV ARM show no abnormality in growth, glucose metabolism, or survival when compared to age and sex-matched controls; in spite of viral replication in their blood and in a wide variety of tissues, most of the cells making growth hormone are comparatively free from replicating virus (OLDSTONE et al. 1985).

Using reassortant viruses between the ARM and the WE strains of LCMV, we mapped the perturbation due to growth hormone deficiency to the S RNA segment of LCMV ARM. C3H/St mice infected with the reassortant (WE/ARM) containing the S RNA segment of the virulent ARM strain showed the manifestations of growth hormone deficiency, including diminished growth, hypoglycemia, and poor survival. Immunochemical studies of growth hormone-containing cells from such mice indicate that the majority express LCMV NP. In contrast, C3H/St mice infected with the reassortant (ARM/WE) containing the L RNA segment of LCMV ARM fail to develop the growth hormone deficiency syndrome and show minimal viral replication in cells containing growth hormone. Thus, only the S RNA of the virulent strain is required for the disease to occur. Previously we had shown that the S RNA segment of LCMV encodes the NP and GP1 and GP2; the combined results point to these products as responsible for the growth hormone perturbation (RIVIERE et al. 1985b).

2.2 Genetic Basis of LCMV Pathogenicity: Virulence in Guinea Pig Maps to the L Segment

Infection of adult guinea pigs with the ARM strain of LCMV leads to a subclinical infection (LEHMANN-GRUBE 1971), in which virus replication seems to be restricted to the spleen. In contrast, infection of adult guinea pigs with the WE strain of LCMV is lethal; all the animals die between days 8 and 12 after infection, and high levels of infectious virus are detected in the sera, lungs, spleen, liver, kidneys, and adrenals. Immunofluorescence studies of cryostat sections of these organs reveal that the viral polypeptides are present. Histologic examination of similar sections by light microscopy shows evidence of necrosis in the spleen and adrenals and less frequently in the liver. In addition, immunosuppression of such guinea pigs infected with LCMV WE does not affect the time of mortality, suggesting that the necrosis observed is primarily due to the virus and not to the immune response against this virus (RIVIERE et al. 1985c). In order to assess the role of the L and S RNA segments in this disease, the two possible reassortant viruses between LCM ARM and LCMV WE were tested for their pathogenicity in guinea pigs. The ARM/WE reassortant was (like the ARM parental strain) avirulent, and all the guinea pigs survived the infection. In contrast, the WE/ARM reassortant was virulent and caused a lethal infection, as did the WE parental strain. These results show that the L RNA segment of WE is required for the mortality following this infection of adult guinea pigs and that the L RNA segment of WE is important for virus replication in vivo. The putative polymerase (L protein) of LCMV is encoded by the L RNA segment (BUCHMEIER et al. 1985). Thus, the differing virulence of the two LCMV strains in guinea pigs may be explained by a difference in the polymerase activities of the parental strains (RIVIERE et al. 1985c). In a previous study, KIRK et al. (1980) generated and characterized intertypic reassortant viruses between the ARM and WE strains of LCMV. Their experience was limited to the WE/ARM genotype only, which they found to be weakly virulent in guinea pigs. Our results contrast with theirs, and are strenghtened by the use of both intertypic reassortants (WE/ARM, ARM/WE) and by similar data with another, independently generated WE/ARM reassortant. Currently we cannot explain the discrepancy between our results and the findings of KIRK et al.; however, it is difficult to rule out the occurrence of spontaneous mutation during the selection of the reassorted clones.

2.3 Induction and Recognition of Virus-Specific, H-2–Restricted Cytotoxic T-Lymphocytes Map to the S RNA Segment of LCMV

Splenic populations of cytotoxic T-lymphocytes (CTL) induced in BALB/c WEHI (H-2^d) mice after primary infection with the ARM strain of LCMV efficiently kill syngeneic H-2^d target cells infected with LCMV ARM. In contrast, these CTL kill LCMV Pasteur (Past)-infected H-2^d targets inefficiently. Past-induced CTL kill H-2^d target cells infected with either LCMV ARM or LCMV Past (AHMED et al. 1984). To investigate the RNA segment involved in CTL induction and recognition, intertypic recombinants between LCMV strains ARM and Past were generated. Two independent clones of each genotype (L RNA/S RNA, i.e., ARM/Past and Past/ARM) were obtained. CTL from splenic populations of mice immunized with LCMV ARM/ARM or Past/ARM efficiently lysed target cells infected with LCMV containing the S RNA segment of ARM (ARM/ARM or Past/ARM), but not the targets infected with LCMV containing the S RNA segment of Past (Past/Past and ARM/Past). In contrast, CTL obtained from mice infected with LCMV Past/Past or ARM/Past killed targets infected with any of the four LCMV genotypes (ARM/ARM, Past/Past, Past/ARM, ARM/Past). In addition H-2^d restricted, LCMV ARM-specific CTL clones preferentially lysed syngeneic target cells that carried the S RNA segment of the LCMV ARM genome. Thus, by manipulating the RNA segments of two LCMV strains, we have shown that the S RNA genome is required for LCMV-specific CTL induction, target recognition, and lytic activity (RIVIERE et al. 1986). Recent studies have localized the GP1, GP2, and the NP genes of LCMV to the S RNA segment (RIVIERE et al. 1985a); other experiments have shown that GP1 is highly variable among the different LCMV strains, whereas GP2 and the NP are conserved (BUCHMEIER 1984, personal communication). Furthermore, GP1 is the predominant viral protein in the virion surface and is expressed on the outer plasma membrane of infected cells (BUCHMEIER et al. 1978; BUCHMEIER and OLDSTONE 1979; BRUNS et al. 1983). Therefore, it is probable that GP1 may be the target for CTL recognition.

3 Genetic Reassortants of LCMV: Unexpected Disease

Recently we noted that LCMV reassortants with the WE/ARM or the Traub/ARM genotype cause a lethal disease following neonatal inoculation of BALB/c WEHI mice, but in contrast, parental strains or reciprocal reassortants do not. This disease is characterized by inhibited growth and death. The pathogenic mechanism is the induction of interferon combined with high virus titers and subsequent liver necrosis (RIVIERE and OLDSTONE 1986). The increased pathogenicity of reassortants derived from less pathogenic or nonpathogenic parents has already been observed with other segmented viruses (SCHOLTISSEK et al. 1979; RUBIN and FIELDS 1980; VALLBRACHT et al. 1980). However, this is the first report concerning the arenavirus family and these experimental results suggest that the outbreak of Lassa fever in Africa may have been related to a natural reassortment in vivo of independent isolates of the Lassa virus, another member of the Arenaviridae (PETERS 1984; JAHRLING et al. 1985). Finally, it should be noted that immunochemical or serological assays would not differentiate the virulent reassortants formed from the avirulent parental strain and, thereby, they remain unnoticed.

4 Summary and Conclusions

Mapping the gene products encoded by the L and the S genomic segments of arenavirus is essential for understanding the mode of viral replication and for defining the molecular basis of pathogenesis. The phenomenon of genetic reassortment has been exploited for this purpose for Pichinde virus (VEZZA et al. 1980; COMPANS et al. 1981; LEUNG et al. 1981; HARNISH et al. 1983) and LCMV (RIVIERE et al. 1985a). It has also been used to define the function of viral genes in LCMV pathogenesis (RIVIERE et al. 1985b, c, 1986). In these studies, the S RNA segment of LCMV has been associated with pathogenicity in C3H/St newborn mice. This RNA segment encodes the two glycoproteins, GP1 and GP2, and the NP; we have not yet defined whether GP1, GP2, or NP singularly or in combination are involved in pathogenicity. This finding was expected, since work with other experimental models of reassortants with reoviruses (FIELDS and GREENE 1982), bunyaviruses (BISHOP and SHOPE 1980) and influenza virus (SCHOLTISSEK 1979) has shown that the viral surface glycoproteins or outer capsid proteins are implicated in the disease. Additionally, other experiments have demonstrated that LCMV-specific CTL induction, target recognition, and lytic activity are associated only with LCMV gene products encoded by the S RNA segment. Although the polymerase of influenza virus was recently shown to be one of the proteins recognized by influenza virus-specific CTL (MOSS and BENNICK, personal communication), it seems likely that this will be an infrequent occurrence with LCMV infection because its putative polymerase (L protein) is encoded by the L RNA segment (BUCHMEIER et al. 1985). Interestingly, the L RNA segment of LCMV WE is important for viral replication in vivo and is associated with fatal acute disease following infection of adult guinea pigs.

Just as the use of reassortant viruses has allowed us to map which RNA segment is associated with a disease, the recent cloning and sequencing of the LCMV genome (SOUTHERN et al. 1984; ROMANOWSKI and BISHOP 1985) should now enable us to identify the structural component(s) of the gene(s) involved in the pathogenicity of LCMV.

Acknowledgements. I thank Drs RAFI AHMED, PETER SOUTHERN, MICHAEL BUCHMEIER for contributions to and critical review of the manuscript, Dr MICHAEL B.A. OLDSTONE for support and KATHY NASIF, NATHALIE FERAUDET for preparation of the manuscript. This research was supported in part by NS-12428, AG-04342, and AI-09484, YR was the recipient of Fogarty International Research Fellowship number 1 F05 TW03304.

References

Ahmed R, Byrne JA, Oldstone MBA (1984) Virus specificity of cytotoxic T lymphocytes generated during acute lymphocytic choriomeningitis virus infection: role of the H-2 region in determining cross-reactivity for different lymphocytic choriomeningitis virus strains. J Virol 51:34–41

Bishop DHL, Shope RE (1980) Bunyaviridae. In: Fraenkel-Conrat H and Wagner RR (eds) Comprehensive virology, Vol. 14: Newly characterized vertebrate viruses, Plenum, New York, pp 1–21

Bruns M, Cihak J, Muller G, Lehmann-Grube F (1983) Lymphocytic choriomeningitis virus: VI. Isolation of a glycoprotein mediating neutralization. Virology 130:247–251

Buchmeier MJ (1984) Antigenic and structural studies on the glycoproteins of lymphocytic choriomeningitis virus. In: Compans RW, Bishop DHL (eds) Segmented Negative Strand Viruses. Arenaviruses, Bunyaviruses, and Orthomyxoviruses. Academic, New York, pp 193–200

Buchmeier MJ, Oldstone MBA (1979) Protein structure of lymphocytic choriomeningitis virus: evidence for a cell-associated precursor of the virion glycopeptides. Virology 99:111–120

Buchmeier MJ, Oldstone MBA (1981) Molecular studies of LCM virus induced immunopathology: development and characterization of monoclonal antibodies to LCM virus. In: Bishop DHL, Compans RW (eds) The replication of negative strand viruses. Elsevier/North Holland, New York, pp 71–77

Buchmeier MJ, Elder JH, Oldstone MBA (1978) Protein structure of lymphocytic choriomeningitis virus: identification of the viral, structural and cell associated polypeptides. Virology 89:133–145

Buchmeier MJ, Welsh R, Dutko F, Oldstone MBA (1980) The virology and immunobiology of lymphocytic choriomeningitis virus infection. Adv Immunol 30:275–331

Buchmeier MJ, Southern P, Parekh B, Oldstone MBA (1985) Molecular and topographic analysis of the glycoproteins of lymphocytic choriomeningitis virus (LCMV). Negative strand virus meeting, Cambridge, September

Compans RW, Bishop DHL (1985) Biochemistry of arenaviruses. Curr Topics Microbiol Immunol 114:153–175

Compans RW, Boersma DP, Cash P, Clerx JPM, Gimenez HB, Kirk WE, Peters CJ, Vezza AC, Bishop DHL (1981) The replication of negative strand viruses. Elsevier, New York

Dutko FJ, Oldstone MBA (1983) Genomic and biological variation among commonly used lymphocytic choriomeningitis virus strains. J Gen Virol 64:1689–1698

Fields BN, Greene MI (1982) Genetic and molecular mechanisms of viral pathogenesis: implications for prevention and treatment. Nature 300:19–23

Harnish DG, Dimock K, Bishop DHL, Rawls WE (1983) Gene mapping in Pichinde virus: assignment of viral polypeptides to genomic L and S RNAs. J Virol 46:638–641

Jahrling PB, Frame JD, Smith SB, Monson MH (1985) Endemic Lassa fever in Liberia: III. Characterization of Lassa virus isolates. Trans R Soc Trop Med Hyg 79:374–377

Kirk WE, Cash P, Peters CJ, Bishop DHL (1980) Formation and characterization of an intertypic lymphocytic choriomeningitis recombinant virus. J Gen Virol 51:213–218

Lehmann-Grube F (1971) Lymphocytic choriomeningitis virus. Virol Monogr 10:1–173

Leung WC, Harnish DG, Ramsingh A, Dimock K, Rawls WE (1981) Gene mapping in Pichinde virus. In: Bishop DHL, Compans RW (eds) The replication of negative strand viruses. Elsevier/North-Holland, New York, pp 51–57

Oldstone MBA, Sinha YA, Blount P, Tishon A, Rodriguez M, von Wedel R, Lampert PW (1982) Virus-induced alterations in homeostasis: alterations in differentiated functions of infected cells in vivo. Science 218:1125–1127

Oldstone MBA, Ahmed R, Buchmeier MJ, Blount P, Tishon A (1985) Perturbation of differentiated functions during viral infection in vivo. I. Relationship of lymphocytic choriomeningitis virus and host strains to growth hormone deficiency. Virology 142:158–174

Palese P (1984) Reassortment continuum. In: Notkins AL, Oldstone MBA (eds) Concepts in Viral Pathogenesis. Vol. II. Springer, Berlin Heidelberg New York Tokyo, pp 144–151

Pedersen IR (1979) Structural components and replication of arenaviruses. Adv Virus Res 24:277–330

Peters CJ (1984) Textbook of human virology. PSG Publishing, Littleton

Riviere Y, Oldstone MBA (1986) Genetic reassortants of lymphocytic choriomeningitis virus: unexpected disease and mechanism of pathogenesis. J Virol 59:363–368

Riviere Y, Ahmed R, Southern PJ, Buchmeier MJ, Dutko F, Oldstone MBA (1985a) The S RNA segment of lymphocytic choriomeningitis virus codes for the nucleoprotein and glycoproteins 1 and 2. J Virol 53:966–968

Riviere Y, Ahmed R, Southern P, Oldstone MBA (1985b) Perturbation of differentiated functions during viral infection in vivo: II. Viral reassortants map growth hormone defect at the S RNA of the lymphocytic choriomeningitis virus genome. Virology 142:175–182

Riviere Y, Ahmed R, Southern PJ, Buchmeier MJ, Oldstone MBA (1985c) Genetic mapping of lymphocytic choriomeningitis virus pathogenicity: virulence in guinea pigs is associated with the L RNA segment. J Virol 55:704–708

Riviere Y, Southern PJ, Ahmed R, Oldstone MBA (1986) Biology of cloned cytotoxic T lymphocytes specific for lymphocytic choriomeningitis virus: V. Recognition is restricted to gene products encoded by the viral S RNA segment. J Immunol 136:304–307

Romanowski V, Bishop DHL (1985) Conserved sequences and coding of two strains of lymphocytic choriomeningitis virus (WE and ARM) and Pichinde arenavirus. Virus Res 2:35–51

Rubin DH, Fields BN (1980) Molecular basis of reovirus virulence: role of the M2 gene. J Exp Med 152:853–868

Scholtissek C (1979) Influenza virus genetics. Adv Genet 20:1–36

Scholtissek C, Vallbracht A, Flehmig B, Rott R (1979) Correlation of pathogenicity and gene constellation of influenza A virus. II. Highly neurovirulent recombinants derived from non-virulent or weakly neurovirulent parent virus strains. Virology 95:492–500

Southern PJ, Blount P, Oldstone MBA (1984) Analysis of persistent virus infection by in situ hybridization to whole mouse sections. Nature 312:555–558

Vallbracht A, Scholtissek C, Flehmig B, Gerth HJ (1980) Recombination of influenza A strains with fowl plague virus can change pneumotropism for mice to a generalized infection with involvement of the central nervous system. Virology 107:452–460

Vezza AC, Cash P, Jahrling P, Eddy G, Bishop DHL (1980) Arenavirus recombination: the formation of recombinants between prototype Pichinde and Pichinde Munchique viruses and evidence that arenavirus S RNA codes for N polypeptide. Virology 106:250–260

Webster RG, Laver WG, Air GM, Schild GS (1982) Genetic and molecular mechanisms of viral pathogenesis: implications for prevention and treatment. Nature 300:19–23

State of Viral Genome and Proteins During Persistent Lymphocytic Choriomeningitis Virus Infection

S.J. Francis, P.J. Southern, A. Valsamakis, and M.B.A. Oldstone

1 Initiation of Persistent Infection In Vivo

Persistent infection of mice can be initiated in utero or at birth with several strains of wild type virus (Traub 1936; Hotchin and Cintis 1958; Buchmeier et al. 1980) or by inoculation of adult immunocompetent mice with variant strain(s) of lymphocytic choriomeningitis virus (LCMV) derived from lymphoid cells of mice persistently infected with wild type virus (Ahmed et al. 1984; Tishon and Oldstone 1986). In all three systems, virus replicates to the highest titers 4–7 days postinfection, usually peaking at 10^7–10^8 plaque-forming units (PFU) of infectious virus (per milliliter serum or per gram tissue). Infectious virus is found in the blood and multiple tissues throughout the animal's normal life span, with the titer dropping until it stabilizes at 10^3–10^5 PFU (per milliliter serum or per gram tissue). The persistent state is associated with a significant diminution of LCMV-specific, major histocompatibility complex (MHC) class I restricted cytotoxic T-lymphocyte (CTL) activity. This decrease in CTL activity accounts for the animal's failure to clear viral materials from its tissues. Similarly, several experimental manipulations designed to impair the immune system (surgical thymectomy, use of genetically athymic mice, irradiation, cyclosporin, or cytotoxin medication) result in the establishment of persistent LCMV infection. Regardless of the mechanism by which persistent infection is initiated, infected mice fail to generate sufficient LCMV-specific, H-2–restricted CTLs to clear the virus. However, reconstitution with LCMV-specific immune cells results in successful clearance of viral materials (Oldstone 1986; Oldstone et al. 1986). Although CTLs specific to LCMV are not generated in adequate

Department of Immunology, Scripps Clinic and Research Foundation, 10666 North Torrey Pines Road, La Jolla, CA 92037, USA

Current Topics in Microbiology and Immunology, Vol. 133

Table 1. Establishment of persistent infection in immunocompetent adult mice inoculated with LCMV ARM wild type or clone 13

Adult mice 6–8 weeks old			CTL assay[b]								Virus persistence[c] or clearance			
Experimental group[a]			BALB/C17 (H-2^d)		MC 57 (H-2^b)		SWR/J (H-2^q)		L-929 (H-2^k)		PFU/ml sera		VNA[d] in Viremia[e]	
Strain	(H-2)	LCMV	WT	Cl 13	WT	Cl 13	WT	Cl 13	WT	Cl 13	AVG	Range	tissues	+/total
BALB/W	dd	WT	59 ± 3^f	56 ± 3	ND	ND	12 ± 3	8 ± 4	nil	<1	2.0×10^2	$(4\times10^2-<50)$	nil	2/20
BALB/W	dd	Cl 13	8 ± 2	9 ± 3	ND	ND	1 ± 0.1	<1	<1	<1	1.3×10^5	$(3\times10^5-2\times10^3)$	+++	20/20
C57BL/6	bb	WT	ND^g	ND	58 ± 3	60 ± 5	ND	ND	10 ± 4	12 ± 1	2.0×10^2	$(4\times10^2-<50)$	nil	1/10
C57BL/6	bb	Cl 13	ND	ND	9 ± 4	10 ± 5	ND	ND	2 ± 1	<1	1.2×10^4	$(2\times10^4-4\times10^3)$	+++	10/10
C57 ob/ob	bb	WT	5 ± 2	2 ± 1	25 ± 5	27 ± 4	ND	ND	ND	ND	<50	$(2\times10^2-<50)$	nil	0/5
C57 ob/ob	bb	Cl 13	4 ± 1	2 ± 1	9 ± 2	15 ± 4	ND	ND	ND	ND	5×10^5	$(3\times10^6-6\times10^4)$	+++	14/14
C57 ob/+	bb	WT	25 ± 4	19 ± 3	48 ± 7	33 ± 6	ND	ND	ND	ND	1.2×10^2	$(3\times10^2-<50)$	nil	0/5
C57 ob/+	bb	Cl 13	2 ± 2	1 ± 1	10 ± 3	12 ± 3	ND	ND	ND	ND	4×10^5	$(6\times10^5-2.5\times10^5)$	+++	12/15
C3H/St	kk	WT	2 ± 1	<1	<1	<1	ND	ND	35 ± 6	34 ± 2	<50		nil	0/8
C3H/St	kk	Cl 13	<1	<1	<1	1 ± 0.3	ND	ND	14 ± 2	12 ± 1	8.1×10^4	$(1\times10^5-7\times10^4)$	+++	8/8
SWR/J	qq	WT	6 ± 3	5 ± 2	ND	ND	63 ± 4	78 ± 3	ND	ND	3.1×10^2	$(6\times10^2-<50)$	nil	2/18
SWR/J	qq	Cl 13	1 ± 0.9	<1	ND	ND	14 ± 1	17 ± 3	ND	ND	8.7×10^4	$(2\times10^5-4\times10^3)$	+++	20/20

[a] Mice were inoculated intravenously with 2×10^6 PFU of either LCMV ARM 1371 wild type (*WT*) or LCMV ARM 1371 clone 13 (*Cl 13*). After 6 (BALB/W, C57BL6, C57 ob/ob, C57 ob/+, SWR/J) or 7 (C3H/ST) days, a suspension of splenic lymphocytes in effector to target ratios of 50:1, 25:1, and 12.5:1 was added to a variety of virus-infected H-2 target cells labeled with ^{51}Cr. Previous experiments showed that the effector cells were Th1.2$^+$, LYT2$^+$, and LT34^{nil} (Oldstone et al. 1986).

[b] Percentage of specific ^{51}Cr released from H-2- and non-H-2-restricted, LCMV-infected target cells in-a 5-h assay.

[c] Viral persistence detected in mice 15–30 days after LCMV inoculation. Infectious virus in individual sera titered on Vero cells (PFU).

[d] LCMV nucleic acid (*VNA*) sequences in tissues detected by whole body sectioning of mice. Tested 15 days after viral infection. Similar results were observed in mice tested 60–180 days after initiating viral infection.

[e] *Viremia* (>50 PFU/ml): number of mice with viremia over total population.

[f] Mean value ±1 SD. Five or more individual BALB/W, C57BL/6, C3H/St, and SWR/J mice tested. Three individual C57 ob/ob or ob/+ mice assayed.

[g] *ND,* not determined.

numbers, CTLs specific for other viruses can frequently be induced in LCMV persistently infected animals, indicating a specific immunologic abnormality. Viral persistence was initiated in several mouse strains by inoculation of newborns (less than 18 h old) with wild type LCMV Armstrong (ARM) strain 1371 clone 53B or inoculation of adult, immunocompetent mice with the lymphotropic variant clone 13 (AHMED et al. 1984; TISHON and OLDSTONE 1986) selected from persistently infected mice inoculated with wild type LCMV ARM clone 53B. The association between generation of the LCMV-specific, H-2–restricted CTL and clearance of virus (expression of PFU and viral nucleic acid sequences) is shown in Table 1. As reported elsewhere, although LCMV H-2–restricted CTL responses are abrogated, persistently infected mice do generate LCMV-specific B-cell responses and produce high titers of antibodies to all the LCMV polypeptides (OLDSTONE and DIXON 1969; BUCHMEIER and OLDSTONE 1978; THOMSEN et al. 1985). Quantitatively, the antibody response is equivalent to or exceeds that made in hyperimmune animals, although the subclass of immunoglobulin produced may be different during the acute or immunizing phase from that during persistent infection (THOMSEN et al. 1985; SALMI, pers. comm.). Whether antibodies are generated to those viral epitopes required for neutralization (PAREKH and BUCHMEIER 1986) is not yet known.

2 Detection of Viral Genes in Whole Mouse Sections During Persistent Infection

During the course of persistent virus infection, viral genomes and the proteins they encode may exist in multiple organs in an infected host. We have developed a procedure that allows efficient and reproducible screening of all tissues in an infected host. This technique detects viral genetic material and viral proteins in whole body sections of infected mice. The procedure is based on the ability of ^{32}P-labeled, single-strand riboprobes to detect either genomic sense or complementary sense viral nucleic acid and of antibodies coupled with [^{125}I] staph A to detect viral proteins (see SOUTHERN et al. 1984; BLOUNT et al. 1986 for details). The expression of LCMV nucleic acid sequences in many tissues following in utero, neonatal, or adult infection leading to persistent virus infection is recorded in Fig. 1. How the virus is transmitted in utero has not been completely resolved. Ovaries become infected, and ova express viral antigens, as noted with a fluorescent antibody technique (MIMS 1966). Thus, vertical transmission via the ovum is possible as long as infection of embryonic cells does not interfere with development; however, infected ova have not been transplanted into uninfected females nor have sperm from persistently infected males been studied.

Viral RNA accumulates in a variety of tissues over the life span of persistently infected mice. Studies of whole body sections from a variety of strains of mice persistently infected with LCMV since birth readily reveal viral RNA 15 days after initiation of infection, but viral RNA is difficult to detect only 5 days after initial infection (Fig. 2). Yet, at day 5 virus titers are at their

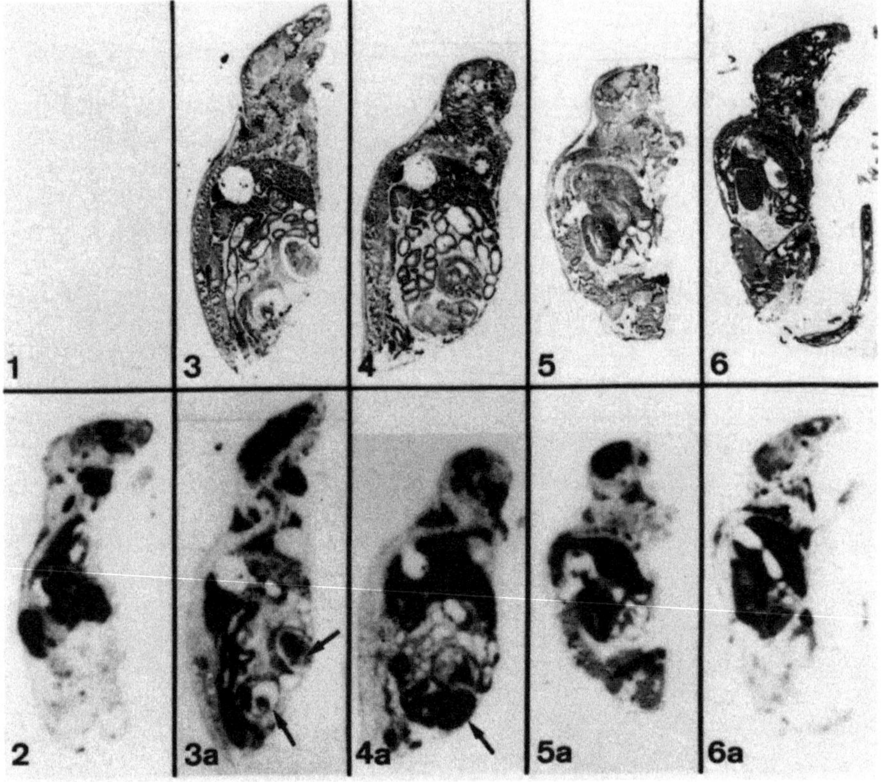

Fig. 1. Expression of LCMV RNA in adult 10–20 week-old mice whose persistent infection was initiated by varying means. Panels *2, 3a,* and *4a* show three mice persistently infected after inoculation with LCMV within 18 h of birth. The persistently infected female mice displayed in panels *3a* and *4a* had been mated with persistently infected males, and infected embryos are visible (*arrows*) during mid (panel *3a*) and late (panel *4a*) pregnancy. These offspring represent the first generation of congenitally infected mice. Panel *5a* pictures the congenitally infected progeny obtained after seven generations of breeding persistently infected parents. Panel *6a* shows a persistently infected mouse that was inoculated at 4 weeks of age with LCMV clone 13 lymphotropic variant. This variant prevents the generation of LCMV-specific H-2 restricted CTLs that are required for clearing virus from infected mice. Hence, panel *6a* represents selective immunosuppression leading to persistent virus infection. For details of the initiation and description of the persistently infected state see the following references: BUCHMEIER et al. 1980; AHMED et al. 1984; TISHON and OLDSTONE 1986. Hybridization was performed with a ^{32}P-labeled LCMV-specific cDNA probe of 200–1600 base pairs with specific activities of $1–10 \times 10^8$ cpm/µg. The mice were etherized, exsanguinated, shaved, and then frozen in blocks of 2.5% carboxy methylcellulose (CMC) by slow immersion into dry ice-ethanol. The CMC blocks were cut at the relevant plane on an LKB cryomicrotome as significantly modified by Haggerty refrigeration company. Sections 40 µm thick were collected on 3M Scotch brand tape and then processed for hybridization. Viral nucleic sequences are seen in the brain, salivary gland, brown fat pad, liver, spleen, kidney, and embryo. Panel *1* shows a control section from an uninfected adult mouse to which the [^{32}P] LCMV cDNA-labeled probe was applied. A similar negative signal was obtained when sections from LCMV persistently infected mice were exposed to radiolabeled (^{32}P) non-LCMV cDNA probe or when a section from an LCMV persistently infected mouse was pretreated with RNase prior to hybridization with the [^{32}P] LCMV cDNA-labeled probe. Details as to the methodology for detection of viral nucleic acid sequences in whole animal body sections and description of the hybridization probes used is provided elsewhere (SOUTHERN et al. 1984; BLOUNT et al. 1986). Histopathology of the identical whole body section used for nucleic acid studies is shown in panels *3, 4, 5,* and *6* which correspond to panels *3a, 4a, 5a,* and *6a,* respectively. Briefly, after performing the hybridization procedure and accompanying washes, sections were stained with hematoxylin-eosin (modified to substitute lithium carbonate for the ammonium carbonate wash) and then exposed to X-ray film

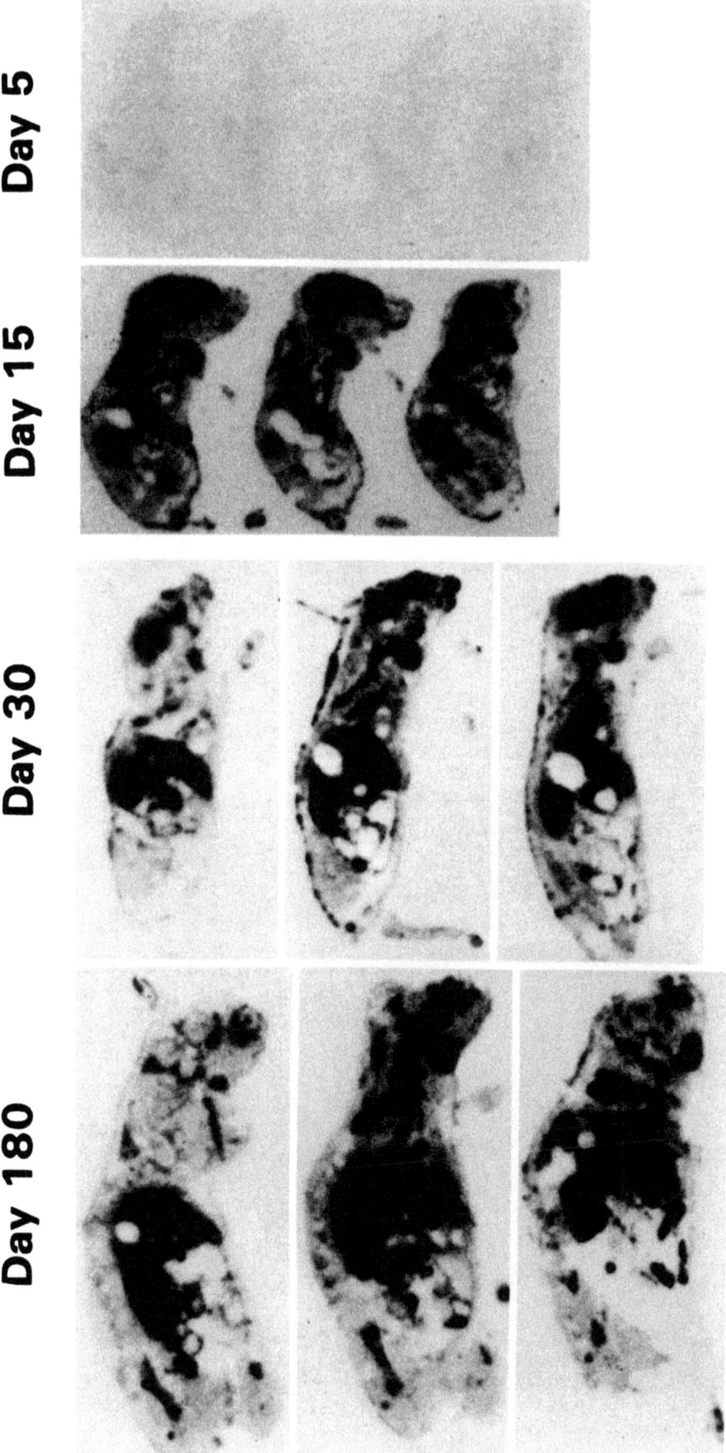

Fig. 2. Increasing expression of viral genetic material with age in persistently infected mice. BALB/WEHI mice were inoculated intracerebrally at birth with LCMV ARM strain (60 PFU) and killed at the indicated times after initiating infection. Whole body sections from individual mice were hybridized with a cloned cDNA probe derived from part of the LCMV S genomic RNA segment (see SOUTHERN et al. 1984, for details). In situ hybridization with sections from different animals is shown at varying time points (four mice at day 5, three mice at days 15, 30, and 180). The data represent over 100 mice studied, of several strains including those on H-2 bb, H-2 dd, and H-2 qq backgrounds. Photomicrograph published in SOUTHERN et al. (1984)

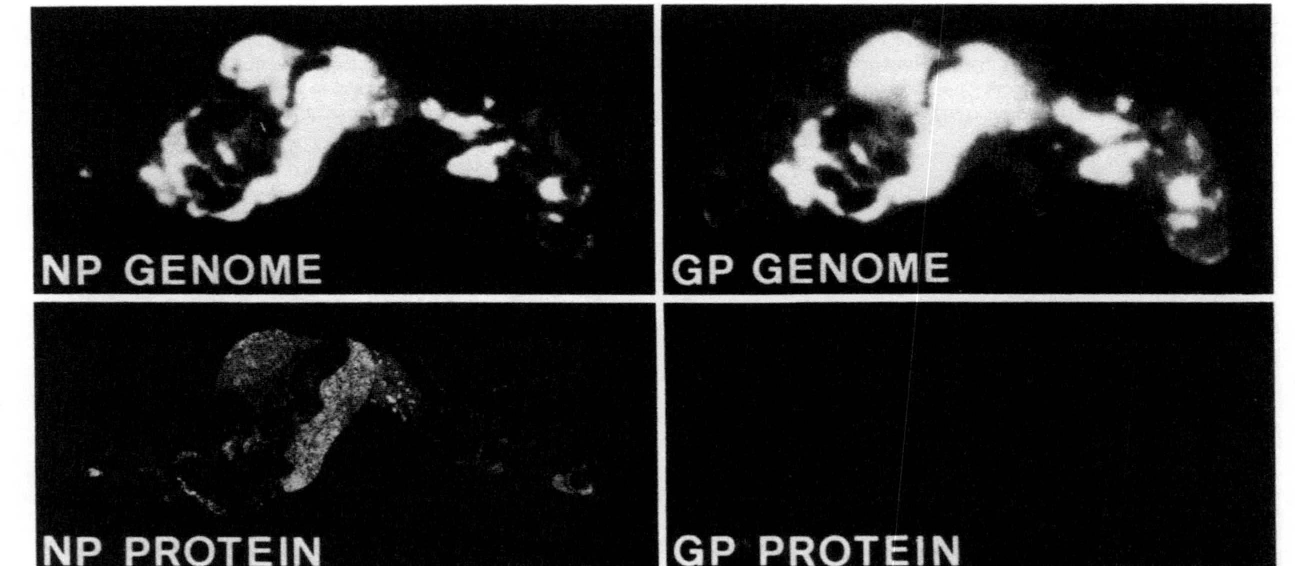

Fig. 3. Expression of LCMV RNA from the nucleoprotein (NP) and glycoprotein (GP) regions and LCMV proteins in a single adult 3-month-old mouse persistently infected with LCMV since birth. LCMV RNA was detected using cDNA probes obtained from the 3′ (NP) and 5′ (GP) ends of the LCMV S (RNA) segment (see BISHOP and AUPERIN, and SOUTHERN and BISHOP, this volume; SOUTHERN et al. 1985). The glycoprotein (GP) m-RNA and the virion S segment RNA are both of genomic sense polarity resulting in an inability to quantitate by in situ hybridization the amount of glycoprotein mRNA present in tissues. Viral nucleic acid sequences are seen in the brain, lacrimal tissues surrounding the orbit, salivary gland, brown fat pad, lung, liver, spleen, kidney, and parts of the intestinal tract. Antibodies specific to either LCMV NP or GP were obtained by immunizing rabbits with amino acid sequences deduced from the open reading frames of the corresponding genes or immunization with purified protein. Expression of LCMV NP was apparent, but there was a restriction in the expression of LCMV GP. Similar results were obtained when monoclonal antibodies directed to either NP or GP were used (for details, see OLDSTONE and BUCHMEIER 1982). The restricted expression of LCMV GP during persistent infection of animals paralleled that seen with a wide variety of persistently infected cultured cells but contrasted to the expression of LCMV GP observed during acute LCMV infection in either cultured cells or animals

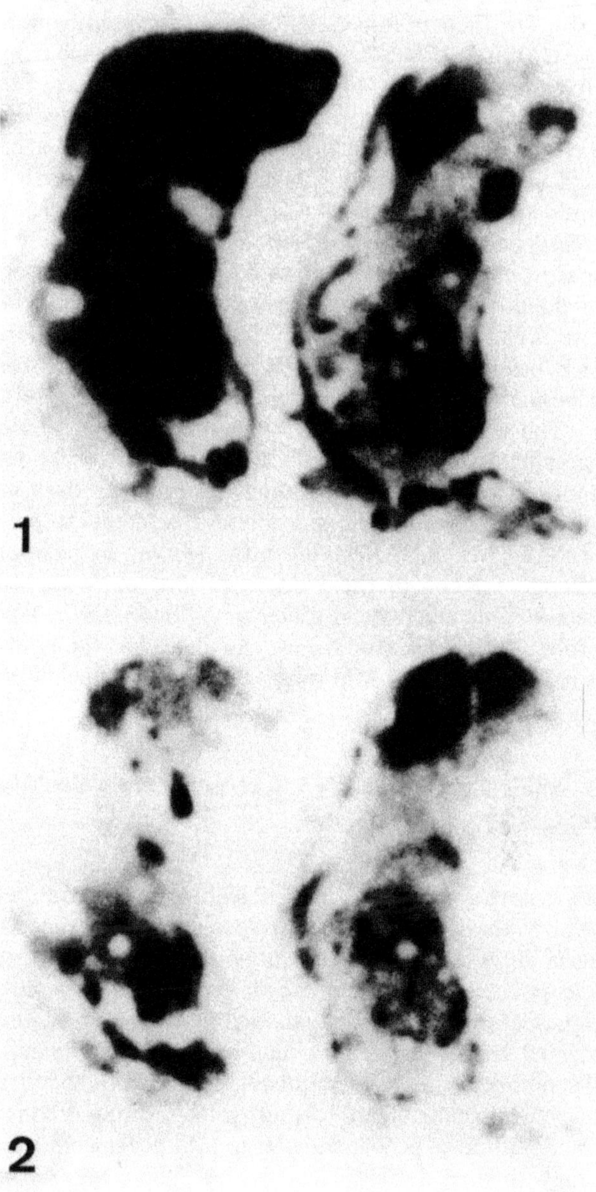

Fig. 4. Expression of genomic sense or complementary sense LCMV RNA in mice persistently infected with LCMV since birth. Single-stranded LCMV riboprobes were made by cloning into SP6 vector. Single-stranded probes were labeled with [^{32}P] UTP and hybridized to two persistently infected mice. The accumulation of genomic sense RNA over that of genomic complementary sense RNA is readily apparent and was a reproducible observation in numerous mice analyzed at different ages postinfection

highest levels and exceed those at 15 days postinfection by 1–2 orders of magnitude. The titers of infectious LCMV recovered from brains of individual BALB/WEHI mice 5 days after initiating infection range from 6×10^6 to 1×10^8 PFU/g. By comparison, viral titers in the brain at 15 days postinfection average 1×10^5 PFU/g and at 180 days are 8×10^3 PFU/g. Nevertheless, such mice after 180 days of persistent infection show an enhanced accumulation of viral nucleic acid sequences (Fig. 2). Thus, during the course of persistent infection, one finds a disparity between the levels of viral nucleic acid sequences, which increase, and the titers of infectious virus, which decrease. With the use of probes specific for the L segment of LCMV (SOUTHERN et al. 1984), the S segment, or the nucleoprotein and glycoprotein genes located on the S segment (RIVIERE et al. 1985; SOUTHERN et al. 1986), we find no difference in the tissue patterns of hybridization (Fig. 3). However, use of riboprobes showed accumulation of genomic sense RNA over complementary sense RNA (Fig. 4).

The sensitivity of the in situ hybridization technique is assessed by dot blot experiments using cloned LCMV cDNAs as standards. First, selected tissues are excised from a whole animal section, and the total nucleic acid is extracted with SDS/proteinase K and phenol treatments to provide a direct measurement of the LCMV RNA content in the section. For example, there are an estimated 250 pg of RNA in the kidney sections of 30-day-old BALB or SWR/J persistently infected mice, as determined by hybridization with an S-specific cDNA probe. From such studies, we calculate that the hybridization of tissue sections is reduced about 10-fold relative to the dot blot method.

3 Analysis of Viral RNA Extracted from the Whole Animal During Persistent Virus Infection

Experiments using the in situ technique to study whole mouse sections have allowed the screening of many sites for the presence of virus proteins and nucleic acids (Figs. 1–4). This technique provides some indication as to the amount and polarity (genomic sense or complementary sense strand) of the sequence representation of individual regions of the virus genome and can be complemented by Northern blot analysis, which supplies additional information on the number, size, and sequence composition of viral RNA species expressed during infection. The Northern blot technique consists of electrophoretic size fractionation of RNA, transfer to a filter that binds nucleic acid, and hybridization with labeled nucleic acid probes. There were two major questions we wished to examine by Northern analysis: (1) is there any alteration in the number or quantity of viral genes expressed during persistent infection when compared to acute infection, and (2) is there any evidence for aberrant transcription or replication of the viral genome during persistent infection?

Figure 5 shows a Northern blot assay with kidney RNA from persistently infected mice. This result is representative of the hybridization patterns seen with other organs. The significant findings are as follows: (1) viral nucleic acid is detectable by Northern blot assay but not by whole mouse in situ hybridiza-

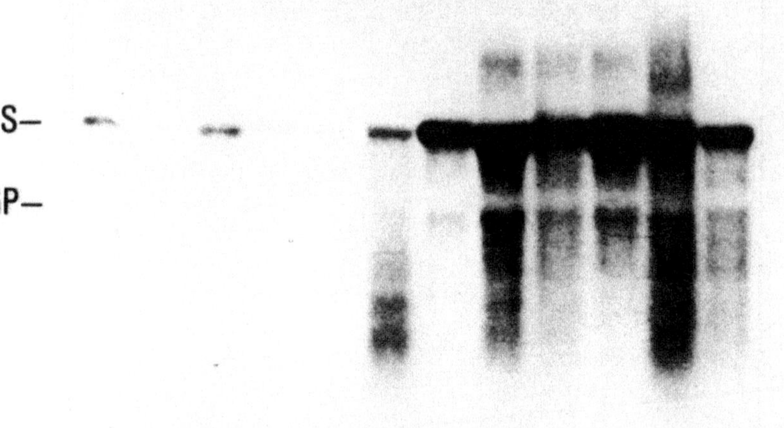

Fig. 5. This figure shows Northern blot assays of kidney RNA samples over the time course of a persistent infection and is generally representative of results seen with other organ RNA samples. RNA from LCMV acutely infected and uninfected BHK cells provide positive and negative control samples for the hybridization probes (*S*, genomic-sized S RNA; *GP*, putative glycoprotein mRNA). The ability to remove radiolabeled probe from the filter and rehybridize with LCMV probes from different regions of the genome allowed meaningful comparisons as to the sequence content of the RNA species identified. To do these studies, newborn mice were infected within the first 24 h of life by intracerebral injection with 60 PFU of LCMV ARM CA 1371. Pairs of animals were sacrificed at the times indicated postinfection and multiple organs (brain, liver, lung, kidney, spleen, and whole blood) were rapidly removed, frozen in liquid nitrogen, and homogenized in guanidinium thiocyanate. RNA from individual tissues was purified by pelleting through a cesium chloride cushion. The RNA was then size fractionated by electrophoresis through a denaturing (formaldehyde or glyoxal) agarose gel, transferred to nitrocellulose, baked, and hybridized with a ^{32}P-labelled, nick-translated, LCMV-specific probe

tion at 5 days postinfection; (2) the total amount of viral nucleic acid increases and remains at a high level at 2 weeks postinfection; (3) genomic RNA is the predominant RNA species present; (4) the presumptive messenger RNA species are relatively abundant; and, most strikingly, (5) there is a diffuse, heterogeneous signal from subgenomic S RNAs. We consider this last finding consis-

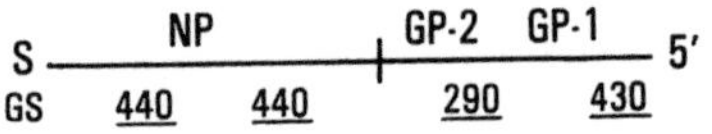

S ————— NP ————|— GP-2 — GP-1 — 5'
GS   440   440   290   430

Fig. 6. A physical map of the S segment of the LCMV genome is displayed. The spatial alignment of probes from regions representing the nucleoprotein and glycoprotein genes are depicted

tent with the following possibilities: in vivo processing of RNA, aberrant transcription, and/or aberrant replication. Artifacts from either RNase contamination or aberrant electrophoresis appear unlikely because we have never seen a similar, heterogeneous hybridization pattern with RNA from acutely or persistently infected cell lines or acutely infected mice. Further, two different RNA isolation procedures have given us similar results, and hybridization to the same filters with a strand-specific riboprobe complementary to actin mRNA reveals no evidence for degradation of actin mRNA. Since selective sparing of actin mRNA by RNase is highly improbable, any nonspecific degradation of RNA during isolation is excluded. We have also electrophoresed and denatured RNA samples under three different conditions and have seen similar hybridization patterns with each technique.

Sequential hybridization of the same filter with LCMV-specific probes from spatially distinct regions of the S genome reveals a pattern of differential hybridization, i.e., probes from the 5' terminus (GP1 region) and from the NP region hybridize to RNA species of smaller molecular size than an internal probe from the GP2 region (Fig. 6). This suggests the presence of deleted or copy back RNAs that may be lacking sequences from the central region of the genomic S segment.

The very complex pattern of LCMV gene expression during persistent infection in vivo requires further characterization. Defective RNAs might be present, especially since interfering or defective interfering (DI) viruses have been described, but not biochemically characterized, in persistent LCMV infections (BUCHMEIER et al. 1980; JACOBSEN and PFAU 1980; POPESCU and LEHMANN-GRUBE 1976). Extensive studies with other viral systems to identify DI particles in vitro have usually, though not always, demonstrated a simple pattern of prominent subgenomic RNA(s) with a marked decrease in the amount of viral genomic RNA (PERRAULT 1981; LAZZARINI et al. 1981; RAO and HUANG 1982). Figure 7 contrasts the viral gene expression in three cell lines persistently infected with LCMV in vitro and three organs taken from a single, 2-month-old, persistently infected BALB/WEHI mouse. The cell lines differ from the organs by their well-defined (but nonidentical) pattern of subgenomic LCMV RNA species. The genomic-sized RNA remains the predominant RNA species both in vitro and in vivo. Therefore, these defective RNAs do not appear to exert any absolute interference in the replication of genomic RNAs.

Replication of the genomic sense strand to produce the genomic complementary sense strand is probably required for subsequent transcription of glycoprotein mRNA (AUPERIN et al. 1984; SOUTHERN and SINGH, pers. comm.; BISHOP and AUPERIN, this volume). We have previously demonstrated a decreased level of glycoprotein during persistent infection (relative to acute infection) both in tissue culture cells (WELSH and BUCHMEIER 1979) and in mice, and have suggested that this may influence immunologically mediated clearance of virus

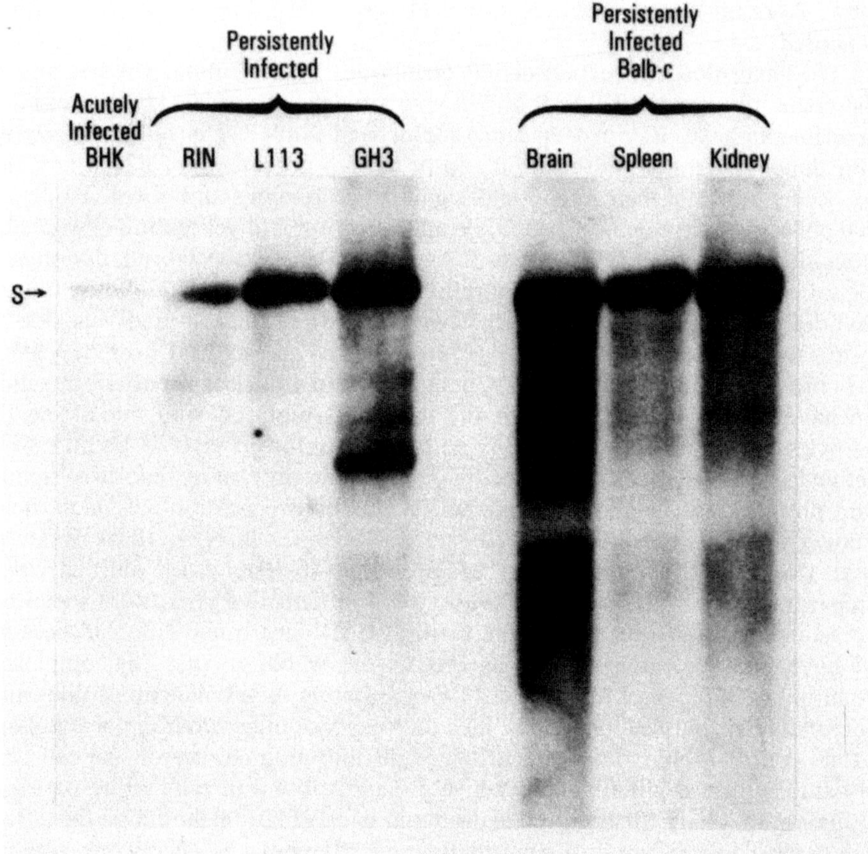

Fig. 7. Northern blot assay of persistently infected tissues and in vitro infected cells. RNA from one acutely infected celline (baby hamster kidney cells, BHK), three cell lines (rat pituitary cells, *GH3;* rat insulinoma cells, *RIN;* B-cell hybridoma, *BHK;* anti-LCMV glycoprotein, *L113*) persistently infected with LCMV, and three organs (*brain, spleen, kidney*) from a single 2-month-old persistently infected BALB/c mouse were size fractionated, transferred, and hybridized with a strand-specific riboprobe (that detects genomic sense RNA) from the nucleoprotein region

(OLDSTONE and BUCHMEIER 1982). The detection of an RNA species corresponding in size to the putative glycoprotein mRNA suggests that regulation of viral glycoprotein synthesis occurs as a posttranscriptional event.

The structure of the in vitro subgenomic RNAs is presently being explored by Northern analysis. Preliminary experiments with strand-specific hybridization probes are indicative of replicative forms because each subgenomic band contains both genomic and genomic complementary sense RNAs. A snapback type of RNA species could give similar results, but this option seems less likely given the size (on denaturing gels) and the positive hybridization with multiple probes that together span the length of the S segment. It appears that defective RNAs generated in vitro with LCMV may have similar structural characteristics (deletions) to DI RNAs previously described in other viral systems (PERRAULT

1981; LAZZARINI et al. 1981; RAO and HUANG 1982), but further information is needed.

We have used two experimental techniques to determine whether any of the tissue subgenomic RNAs found in vivo are deleted species. Hybrid selection experiments have provided evidence for deleted viral RNAs but their relative abundance (aberrant transcription and/or in vivo processing of RNA may also be occurring) and their functional significance remains unknown. HOLLAND and colleagues have demonstrated the amplification and generation of vesicular stomatitis virus (VSV) DI particles in vivo, but in a very specialized, nonphysiological experimental setting (high multiplicity passage in neonatal mouse brains) that did not result in a state of persistent infection (HOLLAND and VILLARREAL 1975; VILLARREAL and HOLLAND 1976). Whereas the natural host of VSV is the cow, LCMV naturally infects rodents, and the model of persistent infection we have used appears similar to the infection found in wild mice. The DI particles generated by VSV appear both in vitro and in vivo to be of a well-defined size by sedimentation velocity, but there are now many reports of significant physical and functional heterogeneity among preparations of DI particles from VSV and other viruses (WEISS et al. 1983; CAVE et al. 1984; BARRETT et al. 1984a, b). It is important to recognize that, in many cases, purified virion preparations were used in which wild type and defective virus were separable by sedimentation. Our experiments with LCMV are quite different because we have analyzed total intracellular RNA from whole organs. The complexity of an organ in terms of multiple cell types, different metabolic and proliferative potential of various cell types, and the multiplicity of microenvironments present differs considerably from a synchronized proliferating culture of one cell type in tissue culture medium. The selective pressures found in cell culture passages at high multiplicity for both replication and encapsidation should be far different, or even nonexistent, in a persistently infected organ – especially one containing a noncytopathic virus such as LCMV. Therefore, the complex pattern of gene expression seen with persistent LCMV infection in vivo is not wholly unexpected.

We have also begun experiments to determine the exact sequence content of some of these deleted RNA molecules by cloning and sequencing. This will allow in vitro manipulation and in vitro generation of deleted RNA molecules for functional testing in transfection assays (LEVIS et al. 1986). These aspects of the investigation are important in order to demonstrate the biological significance of the RNAs described.

4 Detection of Viral Proteins in Whole Animal Sections During Persistent Infection

Using antibodies to predetermined amino acid sequences (SUTCLIFFE et al. 1983), polyclonal, or monoclonal antibodies combined with [^{125}I] staphylococcal protein A, we have detected as little as 5 ng of viral protein per 40 µm section of a whole mouse body and have noted expression of viral proteins in a variety

Table 2. Dissociated expression of LCMV glycoprotein and nucleoprotein during persistent virus infection. Individual neurons were studied for the expression of viral glycoprotein (*GP*) or nucleoprotein (*NP*) by using appropriate monoclonal antibody and antibody to mouse immunoglobulin conjugated to rhodamine or fluorescein isothiocyanate. Sections (4 μm) through the anterior one-third of the temporal lobe were fixed in ether/alcohol (1:1) and 95% alcohol before staining as described elsewhere. At least 100 individual neurons per section were studied. Acute infection was induced in 4–5-week-old mice by intracerebral inoculation of newborn mice with a similar route of inoculation and dose of virus. Infected mice were killed after infection, on the days indicated. Reproduced from OLDSTONE and BUCHMEIER (1982)

Strain		Expression of viral gene products in neurons							
		Acute infection		Persistent infection					
		Day 3–5		Day 5		Day 15		Day >80	
		GP	NP	GP	NP	GP	NP	GP	NP
C3H/St	Adult	8/8[a]	8/8	–[b]	–	–	–		
BALB/W	Adult	5/5	5/5	–	–	–	–		
SWR/J	Adult	8/8	8/8	–	–	–	–		
C3H/St	Newborn	–	–	5/5[a]	5/5	9/15	15/15	1/8	8/8
BALB/W	Newborn	–	–	5/5	5/5	ND	ND	1/20	20/20
SWR/J	Newborn	–	–	5/5	5/5	ND	ND	3/25	25/25

ND, not determined.

[a] Number of mice studied that contained one or more cells expressing a specific viral polypeptide/total number of mice.

[b] The virus dose used in adult mice for acute infection leads to death by day 6–8.

of tissues (BLOUNT et al. 1986). Figure 8 shows the specificity of the protein detection technique as well as the high resolution of viral protein expression obtained on whole animal sections. Figure 9 illustrates that resolution is sufficient to localize LCMV nucleoprotein in several discrete areas of the brain including the cerebellum, cerebral cortex, thalamus, and tissue surrounding the orbit of the eye. Most interestingly, LCMV nucleoprotein has been mapped to the dentate and hippocampal gyri, and within that structure to hippocampal CA1 but not CA4 or CA3 fields (Fig. 9). These results are reproducible when studying the distribution of LCMV proteins in the brains of other persistently infected mice and have been confirmed by immunochemical studies at both the light and electron microscopic levels. The use of monospecific or monoclonal antibodies to the various viral proteins indicates that, whereas viral nucleoprotein is readily expressed, there is a marked reduction in the expression of the viral glycoprotein. This was observed in the whole animal sections reacted with [^{125}I] staph A (Fig. 3) and then using monoclonal antibodies and fluorochrome dyes to look at expression of various viral proteins during acute and persistent infections (Figs. 10, 11; Table 2). Thus, during persistent LCMV infection viral genes are clearly expressed (L, S; GP1, GP2, and NP) (SOUTHERN et al. 1984; Fig. 3), but the degree of viral glycoprotein expression is restricted compared to that of the nucleoprotein (OLDSTONE and BUCHMEIER 1982; Figs. 3, 10, 11;

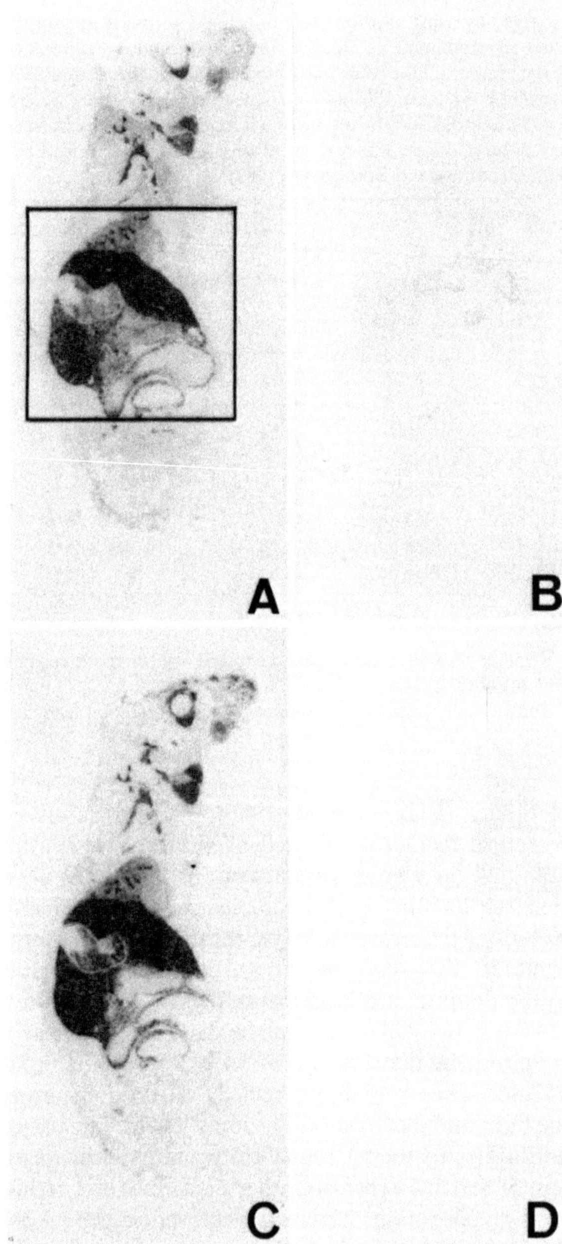

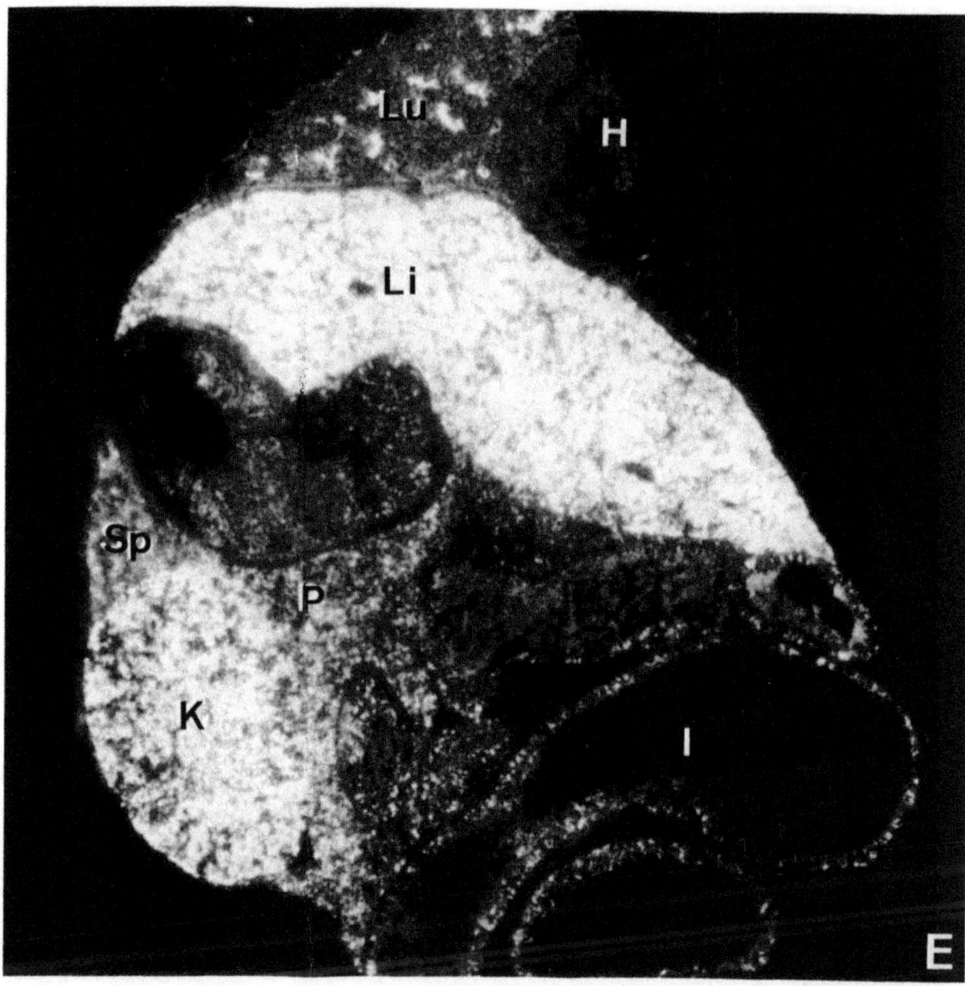

Fig. 8A–E. High resolution of LCMV nucleoprotein expressed in multiple tissue sites of persistently infected animals. Tissues from a 5-month-old CBA/WEHI homologous nude mouse (**A, C**) persistently infected since birth (inoculated with 60 PFU of LCMV ARM strain 1371 clone 53B) and from an uninfected age- and sex-matched control (**B, D**) are displayed. **A** and **C** record staining with a polyclonal guinea pig antibody against LCMV, whereas **C** and **D** demonstrate a rabbit antibody to purified LCMV nucleoprotein. Autoradiography was performed after labeling the antibody bound sections with [^{125}I]staphylococcal protein A. **E** shows a fivefold enlargement of the boxed area in **A,** in which the high resolution and deposition of LCMV nucleoprotein can be observed in a variety of tissues including *L,* lung; *H,* heart; *L,* liver; *S,* spleen; *K,* kidney; *I,* intestinal wall lumen; *P,* pancreas

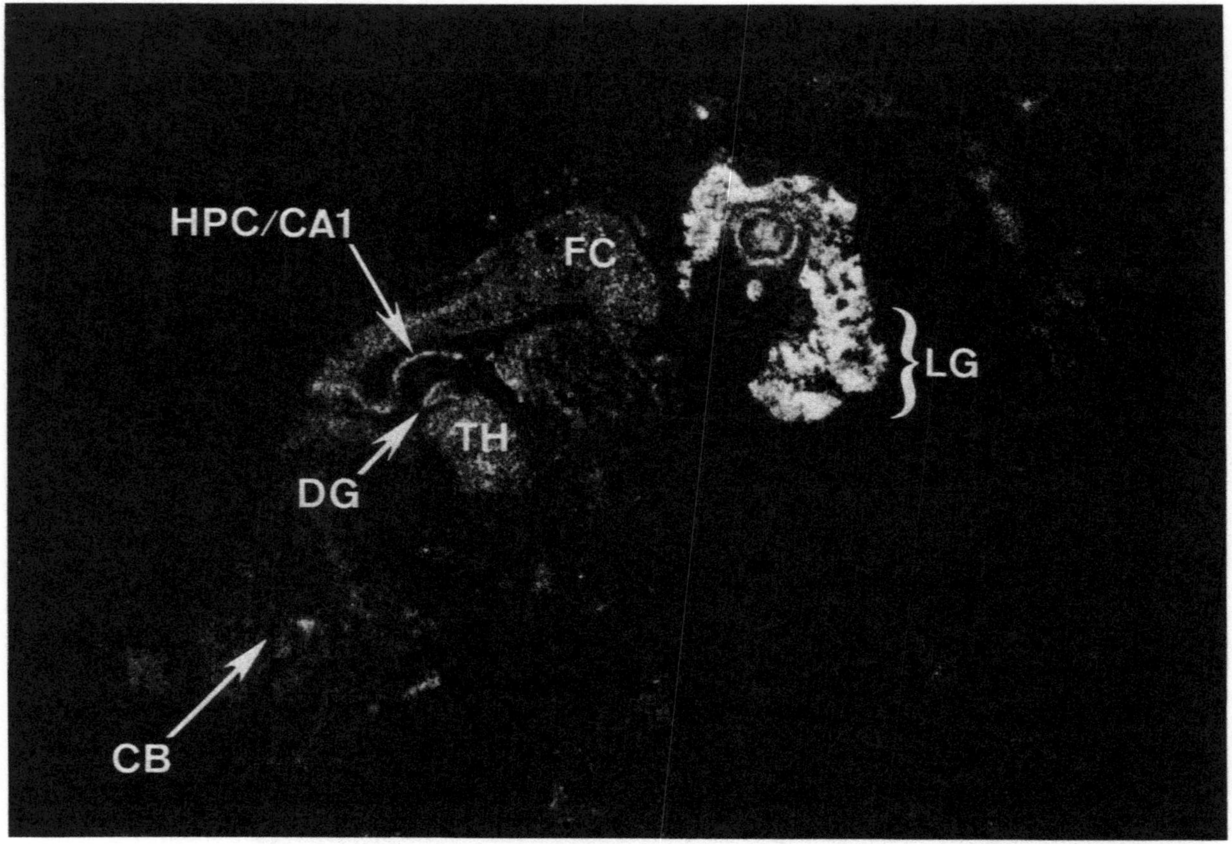

Fig. 9. A 10-fold enlargement of a brain section taken from an adult $3^1/_2$-month-old mouse persistently infected with LCMV since birth. Utilizing rabbit antibody to LCMV nucleoprotein followed by [^{125}I]staphylococcal protein A, the resolution of LCMV nucleoprotein in selected compartments and tracts of the central nervous system are seen. LCMV nucleoprotein is visible in the lacrimal gland tissue around the orbit (*LG*), frontal lobe of the cerebral cortex (*FC*), thalamus (*TH*), cerebellum (*CB*), dentate gyrus (*DG*), and sector CA1 of the hippocampus (*HPC/CA1*). This anatomical distribution of LCMV proteins was routinely seen in over 20 persistently infected mice studied. (See BLOUNT et al. 1986; LIPKIN et al. 1986, for experimental details.) Photo courtesy of W. IAN LIPKIN

ACUTE LCMV (DAY 5) INFECTION OF ADULT MICE

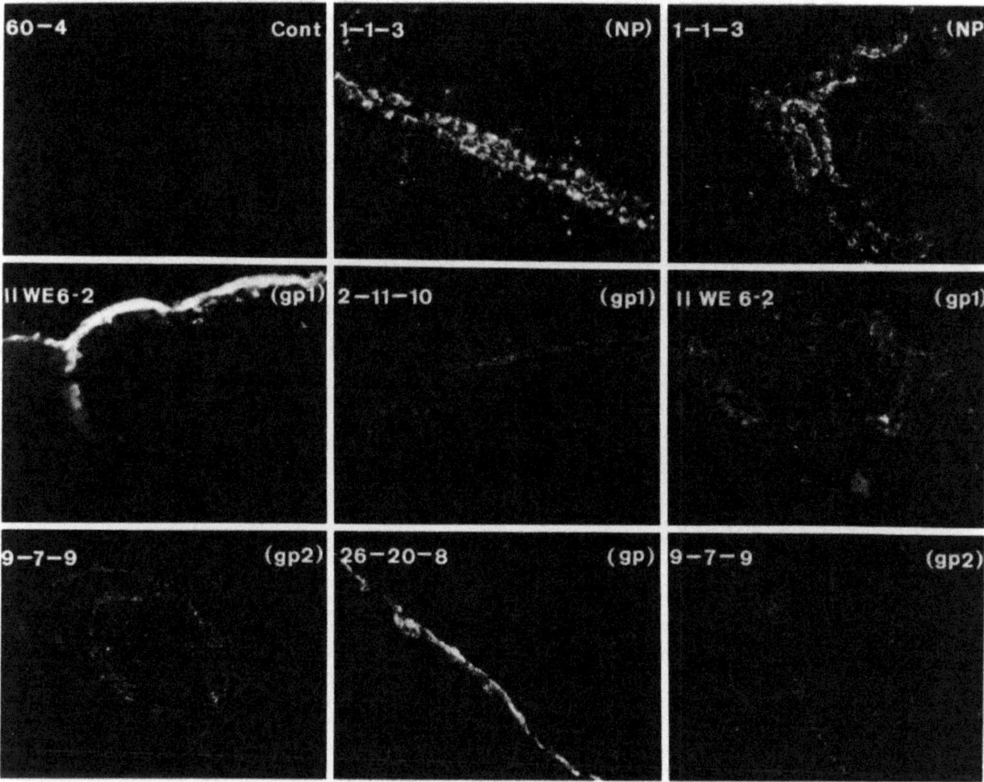

Fig. 10. Expression of viral glycoprotein and nucleoprotein during acute infection of adult mice with 60 PFU of LCMV ARM 1371 strain clone 53B. Expression of both viral glycoprotein and nucleoprotein is evident in cell of the ependyma (*center-upper row*), choroid plexus (*right-upper row; right-middle row; left-lower row*), and meninges (*left and center-middle row; center and right-lower row*). Brains collected 5 days after infection were snap frozen in liquid nitrogen, cut into 4 µm sections in a cryostat and mounted on glass slides. Slides containing the brain sections were fixed and stained, utilizing a variety of monoclonal antibodies. A monoclonal antibody to the glycoprotein (*gp*) of vesicular stomatitis virus, 60-4, served as a negative control. The other monoclonals are listed by their number and the viral component (*NP*, nucleoprotein; *gp1*, glycoprotein 1; *gp2*, glycoprotein 2). After 30 min incubation the cells were washed in buffer and stained with monospecific goat-antimouse immunoglobulin antibody conjugated to rhodamine. The figure represents a photomicrograph of fluorescent microscopic study. Results were observed in BALB/WEHI, C3H/St, C57B1/6, and SWR/J mice (from OLDSTONE and BUCHMEIER 1982)

PERSISTENT LCMV (DAY 84) INFECTION OF ADULT MICE

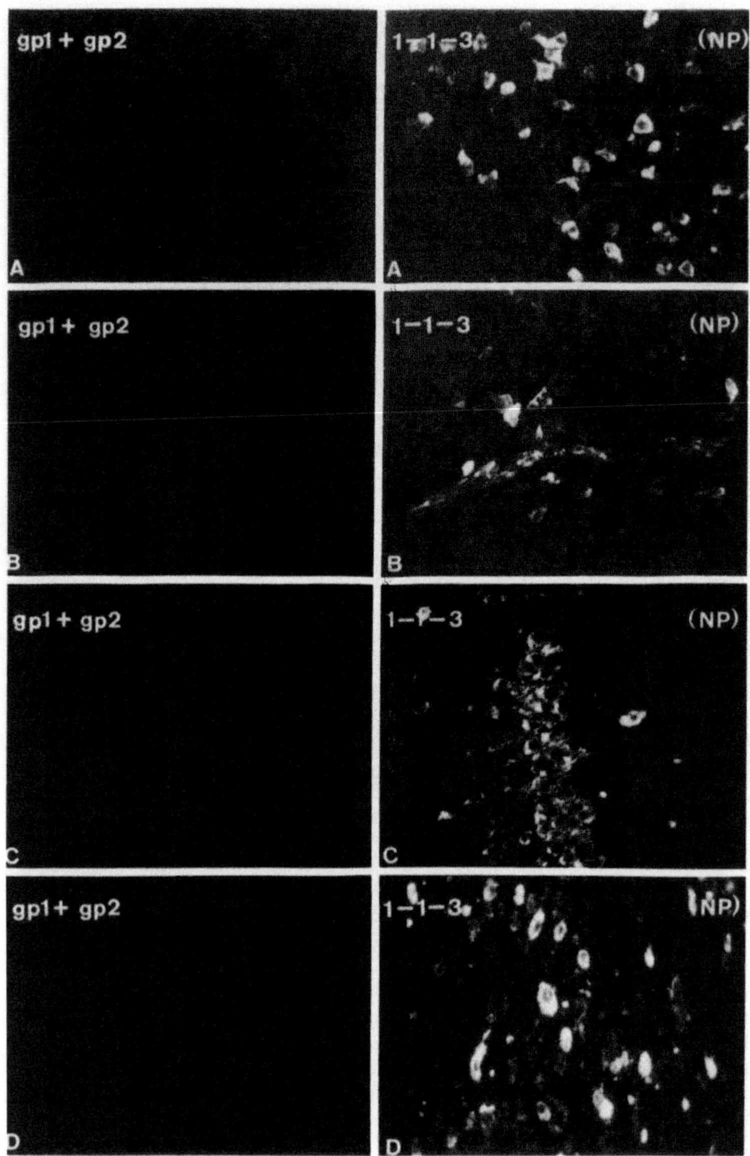

Fig. 11 A–D. Lack of viral glycoprotein but presence of NP in individual neuronal cells from mice persistently infected with LCMV. Newborn mice were inoculated with 60 PFU of LCMV ARM 1371 strain, clone 53B, and killed 84 days later. Neuronal cells of four mice (**A–D**) expressed NP, but in the adjacent 4 µm section no glycoprotein is apparent for the same cells. The collection, sectioning, and staining of the tissues were the same as described in Fig. 8, except that antibodies to GP1 and GP2 from hybridomas were pooled and then used as the staining reagent. Results were observed in BALB/WEHI, C57B1/6, and SWR/J mice or C3H/St mice congenitally infected (from OLDSTONE and BUCHMEIER 1982)

Table 2). This selective decrease in expression of viral glycoprotein is seen in persistent infection but not acute infection (Figs. 10, 11).

Similar results were obtained by WELSH and BUCHMEIER (1979), who studied the expression of viral glycoprotein and nucleoprotein by polyacrylamide gel electrophoresis analysis of [^{35}S]methionine-labeled cells acutely and persistently infected with LCMV. They recorded a 10- to 100-fold decrease in LCMV glycoprotein as compared to the expression of viral nucleoprotein over the course of the infection. Since a virus must escape recognition by the host's immune system to persist in a host cell, the decrease in glycoprotein expression may represent selective gene regulation that provides a mechanism to avoid immunologic surveillance.

5 Consequences of Persistent Infection

At least two immunologically significant consequences mark persistent infection with a noncytolytic virus. Continued exposure to viral proteins combined with an antibody response against the virus results in the formation of virus antigen-antibody complexes. Such immune complexes may deposit in arteries, renal glomeruli, or the choroid plexus, potentially causing arteritis, glomerulonephritis, or choroiditis, respectively. The basic mechanisms and control by which this occurs in arenavirus infection have been analyzed and extended to a variety of RNA and DNA viruses that cause persistent and/or latent infections, as described elsewhere (OLDSTONE 1975, 1984).

Recently, it has been documented that persistent virus infection can alter differentiated physiologic functions of cells without altering cell viability, thereby drastically affecting physiologic functioning of the host (OLDSTONE et al. 1982, 1984a, b). This was first suggested from in vitro studies with LCMV-infected neuroblastoma cells (OLDSTONE et al. 1977) and Rous sarcoma virus-infected chicken chondroblasts, melanoblasts, and muscle cells (HOLTZER et al. 1982) and has been validated in numerous RNA and DNA virus infections of neuronal and lymphoid cells (reviewed in OLDSTONE 1984). This concept of altered specialized function, but not of housekeeping function, with eventual disrupted homeostasis and disease has been supported in vivo during experiments with LCMV persistent infection. In the best studied example, C3H/St mice persistently infected with LCMV ARM strain by inoculation at birth failed to grow normally and developed severe hypoglycemia (OLDSTONE et al. 1982, 1984a, 1985). Multiple lines of evidence indicate that these pathogenic effects are caused by the decreased synthesis of growth hormone in the anterior lobe of the pituitary associated with diminished growth hormone mRNA. Growth hormone mRNA is two- to fivefold less abundant (VALSAMAKIS et al. 1986) than in age- and sex-matched uninfected controls, whereas growth hormone level in the pituitary gland is 50% of the expected normal value (OLDSTONE et al. 1982). Immunochemical analysis by light and high resolution electron microscopy indicates that virus replication is restricted primarily to cells that make growth hormone (OLDSTONE et al. 1982; RODRIGUEZ et al. 1983). When growth hormone levels are reconstituted by adoptively transferring cells that secrete the hormone, the growth retardation and glucose deficiencies of LCMV persistently infected mice are corrected (OLDSTONE et al. 1984). Recent studies have mapped the tropism of LCMV for growth hormone-producing cells of the anterior pituitary to the S genomic RNA segment of the virus (see RIVIERE, this volume; RIVIERE et al. 1985).

It has now become clear that LCMV infection of a unique lymphocyte subset aborts the generation of virus-specific, H-2–restricted CTLs (AHMED et al. 1984). This prevents clearance, and virus persists in tissues of the infected host. During persistence there is an enhanced accumulation of viral nucleic acid sequences, a decrease in production of infectious virus, and a selective decrease in the expression of viral glycoprotein over that of viral nucleoprotein. With fewer of these glycoprotein molecules expressed on the cell surface, there is less opportunity for immune components, humoral or cellular, to recognize infected cells. Thereafter, by virtue of virus tropism for additional differentiated cells, i.e., growth hormone-producing cells, B cells of the islets of Langerhans, etc., normal synthesis of specialized products can be affected, probably at the level of mRNA. The deficiency in a differentiated product upsets homeostatic balance and leads to disease. Not surprisingly, tropisms for many differentiated cells can exist leading to a myriad of physiologic dysfunctions including diseases of the immune, endocrine, or nervous systems.

6   Concluding Remarks

The molecular basis for viral persistence is poorly understood. It is likely that multiple factors including incomplete viral particles or defective interfering viruses, interferon, and generation of mutant viruses all contribute. Analysis of persistent LCMV infection in its natural host, the mouse, has provided considerable information concerning viral gene expression. During LCMV persistence there is a marked accumulation of viral nucleic acid sequences but a drop in the production of infectious virus. Additionally, viral nucleoproteins amass despite a significant reduction of viral glycoproteins. The puzzle of LCMV persistence will begin to unravel as details emerge for the regulation of viral gene expression, for the role of the ambisense genome in virus replication, and for the potential involvement of host cell factors.

Acknowledgements. This is publication number 4361-IMM from the Department of Immunology, Scripps Clinic and Research Foundation. This study was supported in part by USPHS grants AI-09484, NS-12428, and AG-04342. SJF is the recipient of a fellowship from the Juvenile Diabetes Foundation.

References

Ahmed R, Salmi A, Butler LD, Chiller JM, Oldstone MBA (1984) Selection of genetic variants of lymphocytic choriomeningitis virus in spleens of persistently infected mice: role in suppression of cytotoxic T lymphocyte response and viral persistence. J Exp Med 60:521–540

Auperin D, Romanowski V, Galinski M, Bishop DHL (1984) Sequencing studies of Pichinde arenavirus S RNA indicate a novel coding strategy, an ambisense viral S RNA. J Virol 52:897–904

Barrett ADT, Crouch CF, Dimmock NJ (1984a) Defective interfering Semliki Forest virus populations are biologically and physically heterogeneous. J Gen Virol 65:1273–1283

Barrett ADT, Guest AR, Mackenzie A, Dimmock NJ (1984b) Protection of mice infected with a lethal dose of Semliki Forest virus by defective interfering virus: modulation of virus multiplication. J Gen Virol 65:1909–1920

Blount P, Elder J, Lipkin WI, Southern PJ, Buchmeier MJ, Oldstone MBA (1986) Dissecting the molecular anatomy of the nervous system: analysis of RNA and protein expression in whole body sections of laboratory animals. Brain Res (to be published)

Buchmeier MJ, Oldstone MBA (1978) Virus-induced immune complex disease: identification of specific viral antigens and antibodies deposited in complexes during chronic lymphocytic choriomeningitis virus infection. J Immunol 120:1297–1304

Buchmeier MJ, Welsh RM, Dutko FJ, Oldstone MBA (1980) The virology and immunobiology of lymphocytic choriomeningitis virus infection. Adv Immunol 30:275–331

Cave D, Hagen FS, Palma EL, Huang AS (1984) Detection of vesicular stomatitis virus RNA and its defective interfering particles in individual mouse brains. J Virol 50:86–91

Holland J, Villarreal LP (1975) Purification of defective interfering T particles of vesicular stomatitis and rabies viruses generated in vivo in brains of newborn mice. Virology 67:438–449

Holtzer H, Pacifici M, Tapscott S, Bennet G, Payette R, Dlugosz A (1982) Line-ages in cell differentiation and in cell transformation. In: Revoltella RP, Pontieri GM, Basilico C, Rovera G, Gallo RC, Subal-Sharpe (eds) Expression of differentiated functions in cancer cells. Raven, New York

Hotchin JE, Cintis M (1958) Lymphocytic choriomeningitis infection of mice as a model for the study of latent virus infection. Can J Microbiol 4:149–163

Jacobsen SJ, Pfau CJ (1980) Viral pathogenesis and resistance to defective interfering particles. Nature 283:311–313

Lazzarini R, Keene JD, Schubert M (1981) The origins of defective interfering particles of the negative strand RNA viruses. Cell 26:145–154

Levis R, Weiss BG, Tsiang M, Huang H, Schlesinger S (1986) Deletion mapping of Sindbis virus DI RNAs derived from cDNAs defines the sequences essential for replication and packaging. Cell 44:137–145

Mims CA (1966) Immunofluorescence study of the carrier state and mechanism of vertical transmission in lymphocytic choriomeningitis viral infection in mice. J Pathol Bacteriol 91:395–401

Oldstone MBA (1975) Virus neutralization and virus induced immune complex disease: virus-antibody union resulting in immunoprotection or immunologic injury – two sides of the same coin. Prog Med Virol 19:84–119

Oldstone MBA (1984a) Virus can alter cell function without causing cell pathology: disordered function leads to imbalance of homeostasis and disease. In: Notkins AL, Oldstone MBA (eds) Concepts in Viral Pathogenesis, vol. I. Springer-Verlag, Berlin Heidelberg New York Tokyo

Oldstone MBA (1984b) Virus-induced immune complex formation and disease: definition, regulation, importance. In: Notkins AL, Oldstone MBA (eds) Concepts in Viral Pathogenesis, vol. I. Springer-Verlag, Berlin Heidelberg New York Tokyo

Oldstone MBA (1986) Immunotherapy for virus infection. In: Oldstone MBA (ed) Arenaviruses: epidemiology and immunotherapy. Springer, Berlin Heidelberg New York Tokyo (Current topics in microbiology and immunology, vol 134)

Oldstone MBA, Buchmeier MJ (1982) Restricted expression of viral glycoprotein in cells of persistently infected mice. Nature 300:360–362

Oldstone MBA, Dixon FJ (1969) Pathogenesis of chronic disease associated with persistent lymphocytic choriomeningitis viral infection. I. Relationship of antibody production to disease in neonatally infected mice. J Exp Med 129:483–505

Oldstone MBA, Holmstoen J, Welsh RM (1977) Alterations of acetylcholine enzymes in neuroblastoma cells persistently infected with lymphocytic choriomeningitis virus. J Cell Physiol 91:459–472

Oldstone MBA, Sinha YN, Blount P, Tishon A, Rodriguez M, von Wedel R, Lampert PW (1982) Virus-induced alterations in homeostasis: alterations in differentiated functions of infected cells in vivo. Science 218:1125–1127

Oldstone MBA, Rodriguez M, Daughaday WH, Lampert PW (1984a) Viral perturbation of endocrine function: disordered cell function leads to disturbed homeostasis and disease. Nature 307:278–281

Oldstone MBA, Southern P, Rodriguez M, Lampert P (1984b) Virus persists in B cells of islets of Langerhans and is associated with chemical manifestations of diabetes. Science 224:1440–1443

Oldstone MBA, Ahmed R, Buchmeier MJ, Blount P, Tishon A (1985) Perturbation of differentiated functions during viral infection in vivo. I. Relationship of lymphocytic choriomeningitis virus and host strains to growth hormone deficiency. Virology 142:158–174

Oldstone MBA, Blount P, Southern PJ, Lampert PW (1986) Cytoimmunotherapy for persistent virus infection: unique clearance pattern from the central nervous system. Nature 321:239–243

Parekh BS, Buchmeier MJ (1986) Proteins of lymphocytic choriomeningitis virus: antigen topography of the viral glycoproteins. Virology 153:168–178

Perrault J (1981) Origin and replication of defective interfering particles. In: Shatkin AJ (ed) Initiation Signals in Viral Gene Expression. Springer, Berlin Heidelberg New York, pp 151–207 (Current topics in microbiology and immunology, vol 93)

Popescu M, Lehmann-Grube F (1976) Diversity of lymphocytic choriomeningitis virus: variation due to replication of the virus in the mouse. J Gen Virol 30:113–122

Rao DD, Huang AS (1982) Interference among defective interfering particles of vesicular stomatitis virus. J Virol 41:210–221

Riviere Y, Ahmed R, Southern P, Oldstone MBA (1985) Perturbation of differentiated functions during viral infection in vivo. II. Viral reassortants map growth hormone defect to the S RNA of the lymphocytic choriomeningitis virus genome. Virology 142:175–182

Rodriguez M, Buchmeier MJ, Oldstone MBA, Lampert PW (1983) Ultrastructural localization of viral antigens in the CNS of mice persistently infected with lymphocytic choriomeningitis virus (LCMV). Am J Pathol 110:95–100

Southern PJ, Blount P, Oldstone MBA (1984) Analysis of persistent virus infections by in situ hybridization to whole-mouse sections. Nature 312:555–558

Southern PJ, Buchmeier MJ, Ahmed R, Francis SJ, Parekh B, Riviere Y, Singh MK, Oldstone MBA (1986) Molecular pathogenesis of arenavirus infections. In: Brown F, Channock RM, Lerner RA (eds) Vaccines 86:Modern Approaches to Vaccines. Cold Spring Harbor Laboratory Press, New York

Sutcliffe JG, Shinnick TM, Green N, Lerner RA (1983) Antibodies that react with predetermined sites on proteins. Science 219:660–666

Thomsen A, Volkert M, Marker O (1985) Different isotype profiles of virus-specific antibodies in acute and persistent lymphocytic choriomeningitis virus infection in mice. Immunology 55:213–223

Tishon A, Oldstone MBA (1986) Persistent virus infection involving B-cells of islets of Langerhans associated with chemical diabetes: II. Role of viral strains, environmental insult and host genetics. Am J Path (to be published)

Traub E (1936) Persistence of lymphocytic choriomeningitis virus in immune animals and its relation to immunity. J Exp Med 63:847–861

Valsamakis A, Riviere Y, Oldstone MBA (1986) Perturbation of differentiated functions during viral infection in vivo. III. Lymphocytic choriomeningitis virus-induced growth hormone deficiency correlates with decreased growth hormone mRNA in persistently infected mice. Virology (to be published)

Villarreal LP, Holland JJ (1976) RNA synthesis in BHK-21 cells persistently infected with vesicular stomatitis virus and rabies virus. J Gen Virol 33:213–224

Weiss B, Levis R, Schlesinger S (1983) Evolution of virus and defective interfering RNAs in BHK cells persistently infected with Sindbis virus. J Virol 48:676–684

Welsh RM, Buchmeier MJ (1979) Protein analysis of defective interfering lymphocytic choriomeningitis virus and persistently infected cells. Virology 96:503–515

Pathology and Pathogenesis of Arenavirus Infections

D.H. WALKER[1] and F.A. MURPHY[2]

1 Introduction

In systemic virus infections, such as the arenavirus hemorrhagic fevers, pathologic and histopathologic examinations often contribute to the initial understanding of the nature of the disease, but definition of pathogenic mechanisms requires the addition of complementary techniques and experimental approaches – usually in animal models. The commonality of pathologic and histopathologic findings in cases of fatal arenavirus hemorrhagic fever suggests some common pathogenic mechanisms, but the differences reinforce the need for comprehensive study of each disease in naturally infected humans and in experimental animal models that mimic human infections.

[1] Department of Pathology, The Medical School, University of North Carolina at Chapel Hill, Chapel Hill, NC 27514, USA
[2] Division of Viral Diseases, Centers for Disease Control, Atlanta, GA 30333, USA

Current Topics in Microbiology and Immunology, Vol. 133
© Springer-Verlag Berlin·Heidelberg 1987

2 Clinical Spectrum of Human Arenavirus Infections

There are four naturally occurring human arenavirus diseases, lymphocytic choriomeningitis (LCM) and Lassa fever, caused by viruses of the same name, and Argentine and Bolivian hemorrhagic fevers (AHF and BHF) caused by Junin and Machupo viruses, respectively. Much of what we know about these human diseases has been derived from clinical and pathologic investigation of naturally occurring infections, but perhaps more than with most other virus diseases, there has been a major influence from studies of naturally occurring infections in wild host animals that serve as the reservoir and experimentally induced infections in laboratory animals. As in all experimental studies of pathogenesis, care must be taken in extrapolating to human disease mechanisms.

A major portion of the spectrum of interactions between arenaviruses and the infected human host is manifest in what is known clinically as "hemorrhagic fever." However, this clinical term must be complemented by mention of other manifestations of infection. For example, LCMV causes asymptomatic infection, systemic febrile "flu-like" illness, aseptic meningitis, and encephalomyelitis (FARMER and JANEWAY 1942). Fatalities have rarely been documented, although the encephalomyelitis may be lethal, and experimental infections of patients with malignant neoplasms have resulted in death (HORTON et al. 1971). In such patients who develop central nervous system (CNS) disease, the clinical course is bimodal, with an initial "flu-like" illness, fever, headache, and myalgia followed by remission and subsequent return of fever and headache along with other neurologic signs. LCMV has been isolated from blood and cerebrospinal fluid (CSF). The CSF cellular response is primarily lymphocytic; however, no functional or pathologic studies have been done to determine whether the CNS disease in humans results from direct viral pathogenic mechanisms or immunopathologic mechanisms.

Lassa virus also causes a spectrum of clinical manifestations in humans including asymptomatic infection, acute uncomplicated febrile illness, and fatal systemic disease (CASALS and BUCKLEY 1974; FRAME et al. 1970; McCORMICK and JOHNSON 1978; KNOBLOCH et al. 1980). In a study site in Sierra Leone where Lassa virus is endemic, it has been shown that half of all hospital admissions and 30% of deaths in hospitals are caused by Lassa fever. In this setting the mortality rate of these hospitalized patients is 15%–20%; however, in the same setting serosurveys have shown that mild illness, not requiring hospitalization, occurs 10–20 times more frequently than severe illness requiring hospitalization. Further, these surveys have shown that subclinical infection is unusual. Lassa fever is characterized clinically by fever, headache, weakness, myalgia, ulcerative pharyngitis, dysphagia, anorexia, nausea, vomiting, cough, erythema of the face and thorax, facial and cervical edema, rales, rhonchi, stridor, hypotension, hepatic tenderness, pleural effusions, cloudy sensorium, and seizures. Death follows abrupt circulatory collapse and severe hypotension 7–21 days after the onset of illness.

The South American hemorrhagic fevers, AHF and BHF, have a more prominent hemorrhagic diathesis (MOLINAS et al. 1981; VALVERDE-CHINEL 1978). AHF, which has a 10%–15% mortality rate, is characterized clinically by fever, malaise, headache, nausea, anorexia, rash, and hemorrhages, and in some cases CNS involvement starting after about 7 days of illness. After an insidious onset and an initial clinical course similar to that of AHF, BHF patients often deteriorate clinically on days 6–10 with hypotension and/or neurologic signs including tremor, delirium, and seizures. Many of these patients die of circulatory collapse between days 7 and 12. Mild or asymptomatic infections rarely occur.

Overall, the spectrum of clinical manifestations of human arenavirus infections and the spectrum of severity of illness and mortality rate might be taken as indicating an infinite variability in pathologic changes and pathogenic mechanisms, but that is not the case.

3 Pathology of Human Arenavirus Infections

3.1 Pathology of Human Lymphocytic Choriomeningitis

Three necropsies of well-documented human LCMV infection have been reported, one a fatal case of acute meningoencephalitis, and two fatal cases with pneumonia and hemorrhagic manifestations (WARKEL et al. 1973; SMADEL et al. 1942).

The patient with acute meningoencephalitis was a 19-year-old female, who presented with fever, severe headache, confusion, meningismus, and lymphocytic pleocytosis. When examined by brain scan, she had an increased ^{99}Tc uptake, bilaterally, in the parietal lobes. At necropsy, LCMV was isolated from brain tissue, and LCMV antigen was identified immunohistochemically in approximately 15% of cells in the parietal and temporal lobes. Microscopic examination revealed perivascular infiltrates in the meninges, pons, medulla, and Virchow-Robin spaces of the frontal, parietal, and temporal lobes. There was interstitial pneumonia with lymphocytic and mononuclear cell infiltration. Although both LCMV and lymphocytes have been documented in human CNS and CSF, their pathogenic roles in human tissue injury are not known.

3.2 Pathology of Human Lassa Fever

Although Lassa virus causes viremic, panorganotropic, disseminated infection in humans, the number of frankly damaged cells in any organ is insufficient to account for death (WALKER et al. 1982). Principal target organs, as judged by highest virus titers at death, include liver, spleen, lung, kidney, adrenal gland, heart, placenta, and mammary gland, and virus has been recovered from aborted fetal tissues. Macroscopic lesions are seldom observed, although hemorrhagic manifestations such as petechiae and gastric mucosal, renal, and subconjunctival hemorrhages are common. Evidence of increased vascular permeability, such as pleural and pericardial effusions, ascites, and pulmonary, facial, intestinal, and laryngeal edema, have been encountered occasionally.

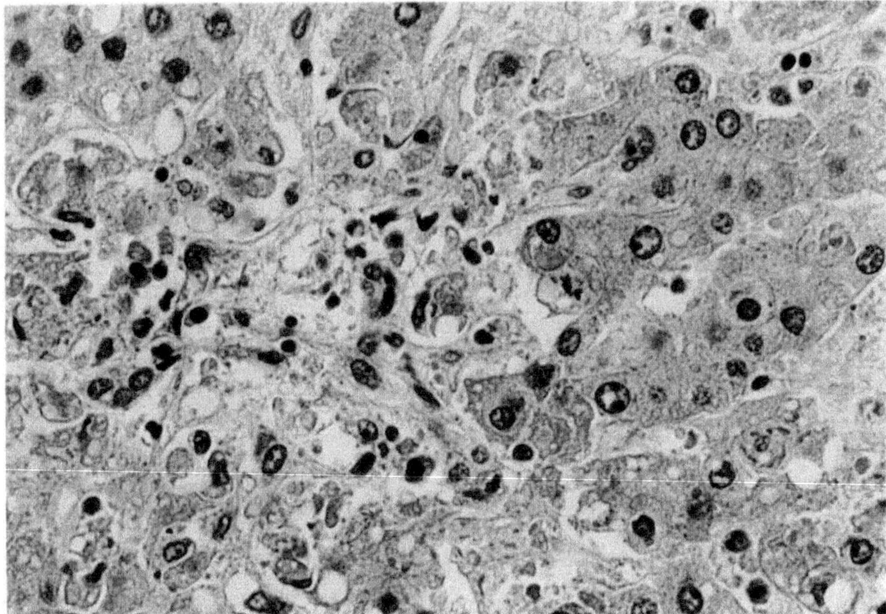

Fig. 1. Human Lassa fever. Liver; focal hepatocellular necrosis with cytoplasmic eosinophilia, nuclear pycnosis, and cytolysis. H&E

Microscopic lesions are widespread throughout the body, although as stated above, at a rather unimpressive magnitude in any given organ or tissue. Lesions are distributed in a characteristic pattern involving liver, spleen, adrenal gland, kidney, heart, lung, and skeletal muscle (FRAME et al. 1970; EDINGTON and WHITE 1972; SARRATT et al. 1972; WINN and WALKER 1975; WALKER et al. 1982; KNOBLOCH et al. 1980). In these organs and tissues cell necrosis and immunohistochemical evidence of cellular infection are focally dispersed and not concentrated at functionally vital sites. Microscopic hemorrhagic diathesis is rare, and thromboses are not evident.

The most consistent microscopic lesion in Lassa fever is multifocal hepatocellular necrosis (Fig. 1). This lesion has been observed consistently from the initial investigations of Lassa fever and was apparent in each of a series of 31 cases with confirmed virologic diagnosis (ISHAK et al. 1982; McCORMICK et al. 1986). The proportion of necrotic hepatocytes in fatal cases has varied from less than 1% up to 50%. Hepatocytes undergo coagulative necrosis or, alternatively, eosinophilic condensation (with Councilman-like bodies), or cytoplasmic vacuolation and rarifaction. Nuclear changes are also variable, with either nuclear pyknosis or lysis. There is usually little inflammatory cellular response associated with this focal necrosis; at best there is a modest presence of macrophages, possibly including activated resident Kupffer cells, at sites of hepatocellular damage. These mononuclear phagocytes engage in engulfment of cellular debris, but their small numbers result in excessive in situ buildup of necrotic cells. Other, less common, histopathologic evidence of hepatocellular injury includes focal cytoplasmic degeneration and steatosis.

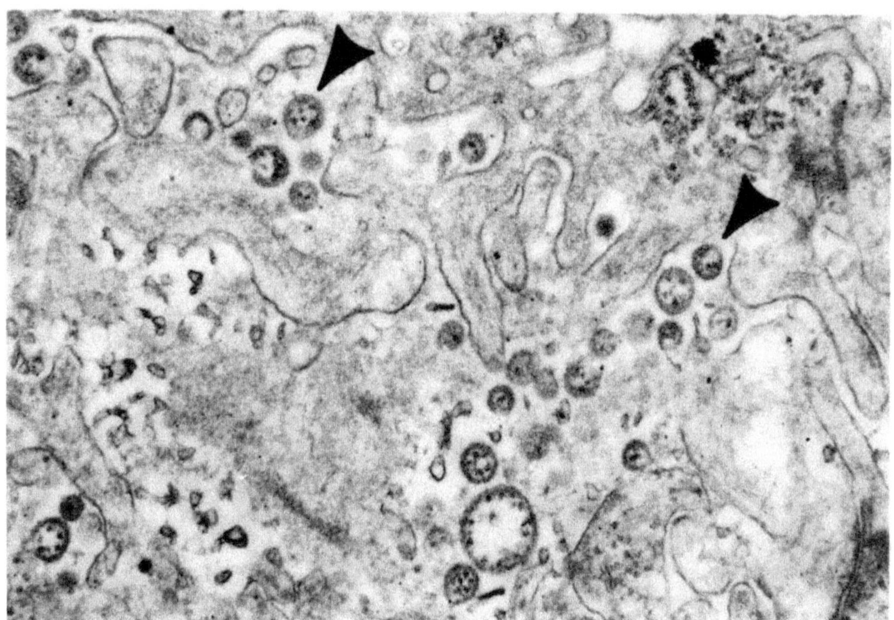

Fig. 2. Human Lassa fever. Liver; typical virions in the extracellular space at the margin of an infected hepatocyte. Thin-section electron microscopy. *Arrows* indicate virions

In ultrastructural studies of liver specimens from 19 patients with Lassa fever, typical arenavirus virions have been demonstrated in only 4 (WINN et al. 1975; McCORMICK et al. 1986). In a case report of an immediate postmortem hepatic biopsy, the ultrastructural observations were particularly instructive (WINN et al. 1975) (Figs. 2–4). In this biopsy tissue, Lassa virions budded from and were associated solely with hepatocytes. Mature virions were located only in the extracellular space, as expected of a virus that is formed by budding from the plasma membrane of the infected host cell. Virions were present in perisinusoidal spaces (the space of Disse), intercellular spaces between hepatocytes, and bile canaliculi formed via lateral hepatocellular junctions. Most virions were associated with less-damaged hepatocytes. Evidence of cell injury included focal cytoplasmic degeneration, dilatation of rough endoplasmic reticulum, flocculent, electron-dense cytoplasmic and mitochondrial deposits, lipid accumulation, marginated chromatin, and end stage necrosis.

This coincidence of hepatocellular infection and damage in the same foci in the liver was also demonstrated in another study in which Lassa virus was isolated from hepatic tissues in six of ten fatal cases. In this study hepatic necrosis was observed in all cases, and alanine aminotransferase was elevated in all eight patients tested (McCORMICK et al. 1986).

From a compilation of histopathologic studies of persons with fatal Lassa fever, the hepatic lesions may be grouped into three categories: (1) lesions marked by the presence of focal cytoplasmic degeneration and necrosis involving less than 20% of hepatocytes, (2) lesions marked by the presence of multifocal necrosis, often with some cells in end stage necrosis, involving 20%–50% of

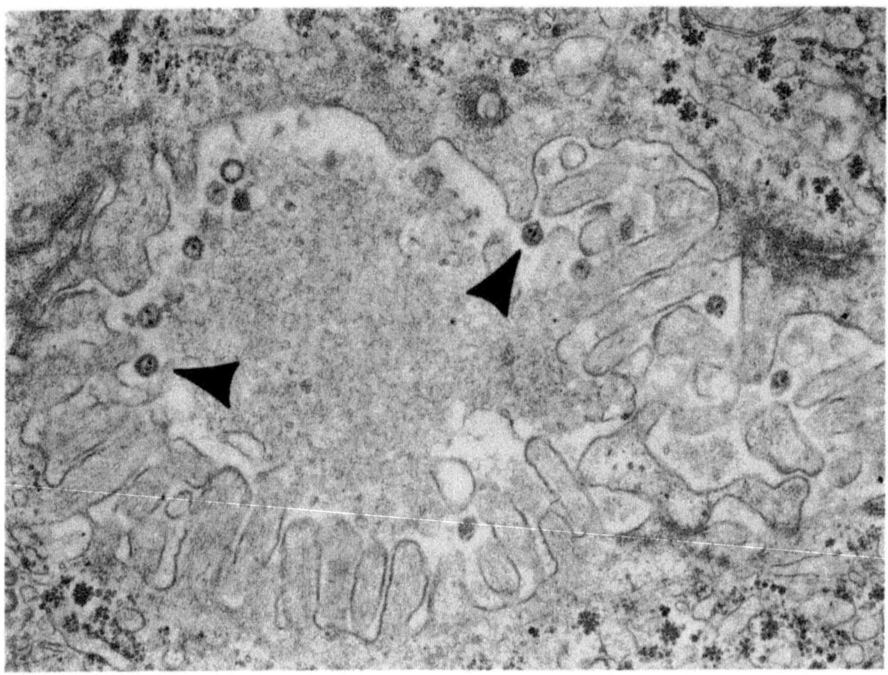

Fig. 3. Human Lassa fever. Liver; virions in bile canaliculus between infected hepatocytes. Thin-section electron microscopy. *Arrows* indicate virions

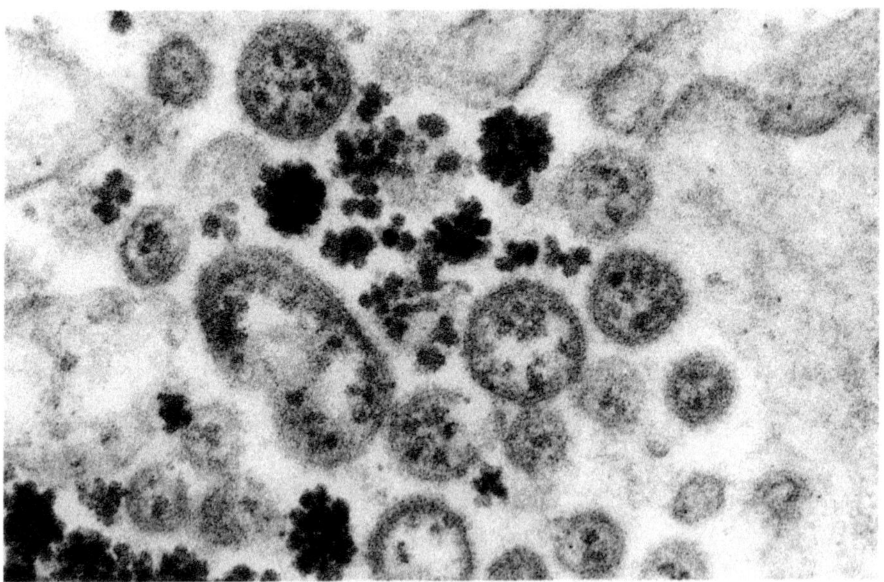

Fig. 4. Human Lassa fever. Liver; high magnification of typical arenavirus virions composed of a membrane envelope surrounding an interior in which there are varying numbers of host cell ribosomes. Thin-section electron microscopy

hepatocytes, and (3) lesions marked by the presence of mitoses, no new focal cytoplasmic degeneration, and necrosis involving less than 10% of hepatocytes. The first of these categories represents patients who died with mild-to-moderate hepatic damage, with lethal damage that might be considered to be centered in other organs and tissues. The second category represents patients with severe Lassa hepatitis, with lethal damage that might be centered in the liver. The third category represents patients with some evidence of continuing damage but also with evidence of regeneration, with lethal damage that must also be considered to be centered in other organs and tissues.

There is a correlation between high virus titers in liver tissue and severe hepatitis – the second category above. In a study of this correlation (McCor-mick et al. 1986), virus titers in severe hepatitis cases were as high as 10^9 $TCID_{50}/g$ of liver tissue, as compared to titers as high as 10^6 $TCID_{50}/g$ in individuals with less than 10% of hepatocytes exhibiting necrosis. Moreover, nearly all of those with severe hepatitis had titers higher than the highest one seen in persons with less severe hepatitis.

Replication of Lassa virus to high titer and concordant severe hepatitis have both been associated with a high probability of fatal outcome. This conclusion is based not only upon cumulative clinical, virologic, and pathologic data, but also upon a prospective clinical study conducted in Sierra Leone (Johnson 1982). In this study mean peak viremia titers in survivors and in fatalities were $10^{2.1}$ and $10^{4.1}$ $TCID_{50}/ml$ of blood, respectively. The mean serum aspartate aminotransferase concentration in survivors was 171 U/liter and in those who died 923 U/liter. Among patients with a serum aspartate aminotransferase concentration of less than 150 U/liter and viremia of less than 10^4 $TCID_{50}/ml$, 88% survived. Conversely, 94% of patients with a serum aspartate aminotransferase concentration of greater than 150 U/liter and a viremia of greater than 10^4 $TCID_{50}/ml$ died. A higher serum concentration of aspartate aminotransferase than alanine aminotransferase suggests that injury to cells other than hepatocytes was crucial.

It is reasonable to hypothesize from the observations cited above that Lassa virus damages hepatocytes directly, as a consequence of acute infection. The observed necrosis is not associated with the accumulation of lymphocytes, monocytes/macrophages, or neutrophils, and at the time of damage there is no evidence of any antibody response to the infection.

Extrahepatic lesions in fatal Lassa fever include splenic necrosis, necrosis of renal tubular cells, focal renal interstitial lymphocytic infiltrates, multifocal adrenocortical necrosis, mild interstitial myocarditis, interstitial pneumonia, and rhabdomyositis (Walker et al. 1982; Edington and White 1972; Winn and Walker 1976; Frame et al. 1970; Knobloch et al. 1980). Splenic necrosis is centered in the marginal zone of the periarteriolar lymphatic sheath and is accompanied by deposition of fibrin. This lesion is virtually constant and reflects viral tropism for reticuloendothelial cells. Multifocal adrenocortical necrosis reflects viral tropism for secretory epithelium and is often associated with a focal inflammatory response in this organ. Renal tubular lesions are few in number and small in size. These lesions are characterized by focal coagulative necrosis, with resulting residual squamous basal cells lining tubules, and multinucleated giant cells intermixed with regenerating tubular epithelium. These renal lesions may represent damage due to a combination of Lassa virus infection, shock, and old injury. Clinical azotemia is likely to be caused by a reduced glomerular filtration rate and occasionally by acute tubular necrosis associated with profound shock.

Mild interstitial pneumonia, capillary congestion, and alveolar edema accompany Lassa virus infection in the lungs; chest roentgenograms are usually normal. Neither myocarditis nor the minimal-to-mild rhabdomyositis observed seems capable of accounting for the elevated concentrations of serum creatine kinase that are striking in some cases. The absence of fibrin thrombi correlates with generally normal functional measurements of coagulation mechanisms; platelet counts are usually normal and do not fall below 50000/µl. It may be concluded that altered hemostasis is not an important factor in the pathogenesis of Lassa fever. Hepatic failure and disseminated intravascular coagulation, likewise, are not involved.

As reasonable hypotheses to explain the pathogenesis of Lassa fever are discarded by the accumulation of clinical, clinicopathological, and experimental data, new hypotheses are called for. In this regard, it is necessary to consider the interactions of leukotrienes, prostaglandins, complement, and kallikrein in the pathogenesis of the sudden vascular collapse that characterizes fatal Lassa fever.

3.3 Pathology of Argentine Hemorrhagic Fever

Hemorrhagic diathesis occurs frequently in the South American hemorrhagic fevers. In a study of 32 patients with AHF, oral and axillary petechiae were noted in 32, hematuria in 18, gastrointestinal hemorrhage in 5, metrorrhagia in 5, gingival hemorrhage in 5, epistaxis in 4, and hemoptysis in 1 (MOLINAS et al. 1981). In these cases there was no correlation between the extent of hemorrhage and the severity of the disease.

In a series of 12 complete autopsies of patients with fatal AHF, the following distribution of lesions was noted: disseminated hemorrhages in 11, myocarditis in 4, hepatocellular necrosis in 8, papillary necrosis or hemorrhage in 3, encephalitis with perivascular lymphocytic cuffs in 3, gastrointestinal ulcers in 8, and pulmonary bacterial superinfection in 7. Hemorrhages involved the lungs in 7, pericardium in 4, renal capsule or pelvis in 5, adrenal glands in 3, CNS in 5, gastrointestinal tract in 5, and hepatic capsule in 4 (ELSNER et al. 1973). Fibrin thrombi were identified in only 3 cases.

In AHF the sites of cellular necrosis have been shown to correspond to sites of viral antigen accumulation, as detected by immunofluorescence. In a study of the tissues of seven victims of AHF, viral antigen was found in the hepatocytes, renal tubular epithelium, macrophages, and dendritic reticular cells of the spleen and lymph nodes (COSSIO et al. 1975; GONZALEZ et al. 1980; MAIZTEGUI et al. 1975). Tissues from the same patients were examined by light microscopy; there was multifocal necrosis in the splenic red pulp and depletion of dendritic reticular cells, as well as depletion of cortical lymphocytes in lymph nodes. Junin virus was isolated from spleen and lymph nodes in all seven necropsies investigated. In some cases comparison of viral titers in spleen, lymph nodes, and blood indicated that viral replication occurred in these lymphoreticular tissues. Likewise, Junin virions were demonstrated by electron microscopy in lymph nodes and spleen.

Other immunohistochemical and clinical laboratory data have shed little light on the pathogenic mechanisms operative in AHF. The absence of detectable deposits of immunoglobulin and complement in renal glomeruli has been taken as strong evidence against the involvement of circulating immune complexes (COSSIO et al. 1975). The lack of correlation of specific coagulation abnormalities with the severity of disease has been interpreted as evidence against the central involvement of a coagulopathy. Similarly, evidence has been accumulating that disseminated intravascular coagulation is not an important pathogenic phenomenon in the disease (MOLINAS et al. 1981; MOLINAS and MAIZTEGUI 1981). Prolonged partial thromboplastin time and low factor VIII:C to factor VIII-related antigen ratio have been noted during an illness, but these parameters return to normal later in its course. Factor V is uniformly elevated; fibrinogen is normal in mild cases and elevated in severe cases. Fibrin degradation products have generally not been detected. Evidence for endothelial injury, including thrombocytopenia, has been found in severe disease, and amounts of factor VIII-related antigen, which is synthesized and released from endothelial cells, are elevated throughout the illness.

3.4 Pathology of Bolivian Hemorrhagic Fever

Necropsies of eight patients with BHF, whose immediate cause of death was shock, have been studied and described; the most common macroscopic lesion observed was gastrointestinal hemorrhage, along with focal ecchymoses and petechiae in many sites. In four of the eight patients there was intracranial hemorrhage (CHILD et al. 1967). Histopathologic lesions included interstitial pneumonia and hepatocellular necrosis. One patient had perivascular lymphocytic cuffs in the CNS, and two had renal tubular necrosis.

In this review of the pathology of human arenavirus diseases there are recurrent themes – hepatocellular necrosis, interstitial pneumonia, hemorrhagic diathesis, shock, CNS involvement, and a significant mortality rate (the exception being LCMV disease), as well as reticuloendothelial tissue infection and damage. Nevertheless, as stated above, it is prudent to consider the pathogenesis of these diseases separately. Generalizations are less likely to provide insight into pathogenic mechanisms than are careful clinical and experimental studies of each disease.

4 Arenavirus-Host Interactions

Over many years research on arenavirus infections in experimental animals, particularly LCMV infection in mice, has led to the discovery of many fundamental phenomena involving and affecting virus-host interactions. These may be divided into four categories: (1) infection is established but is noncytopathic,

(2) there is noncytopathic infection but alteration of an important cellular function, (3) there is a host-mediated immunopathologic consequence, and (4) there is direct virally induced cellular damage. Each of these interactions has been studied in animal models involving a particular species of animal (often an inbred strain) and a particular arenavirus (often a particular laboratory strain of virus). The goal of these studies, besides contributing to the advance of basic biomedical science, has been to further our understanding of the human arenavirus diseases. In the following sections, we attempt to tie each category of virus-host interaction to human disease.

4.1 Noncytopathic Arenavirus Infections

Noncytopathic arenavirus infections are best exemplified by neonatal infections in rodents that are the natural reservoir. Examples of natural rodent-virus interactions include LCM infection in *Mus musculus*, Junin virus in *Calomys callosus*, Machupo virus in *Calomys musculinus*, Tamiami virus in *Sigmodon hispidus*, and Lassa virus in *Mastomys natalensis* (MURPHY and WALKER 1978; WALKER et al. 1976; WEBB et al. 1975; MURPHY et al. 1976) (Figs. 5–8). These infections are generally characterized as persistent with chronic virus shedding and vertical transmission of virus from one generation to the next. Some of these infections meet the precise definition of the term "persistent-tolerant infection", whereas others fail to because of minimal residual host immunological responsiveness to the presence of the virus.

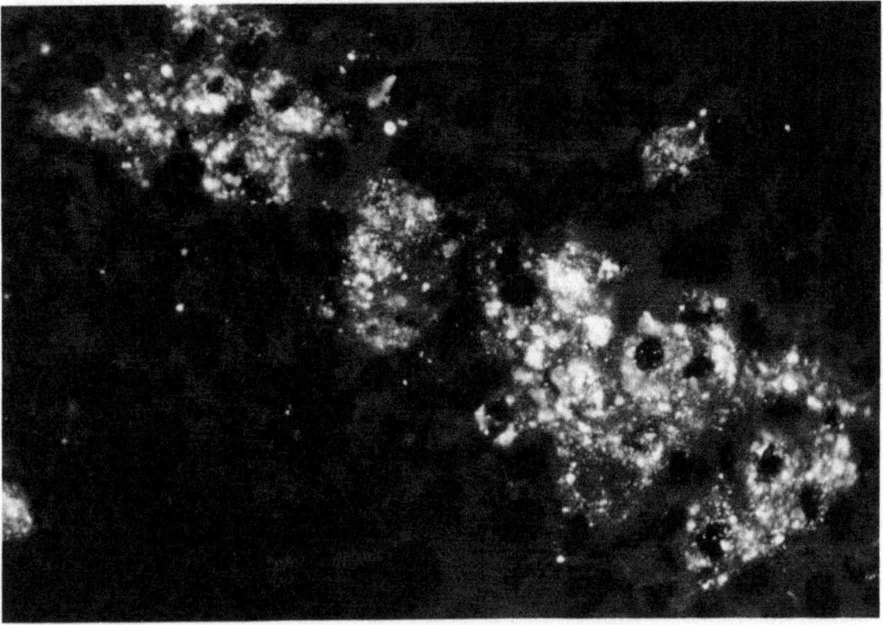

Fig. 5. Tamiami virus infection of its natural rodent host, *Sigmodon hispidus*. Liver; focal hepatocellular infection without cytopathology. Frozen section, immunofluorescence

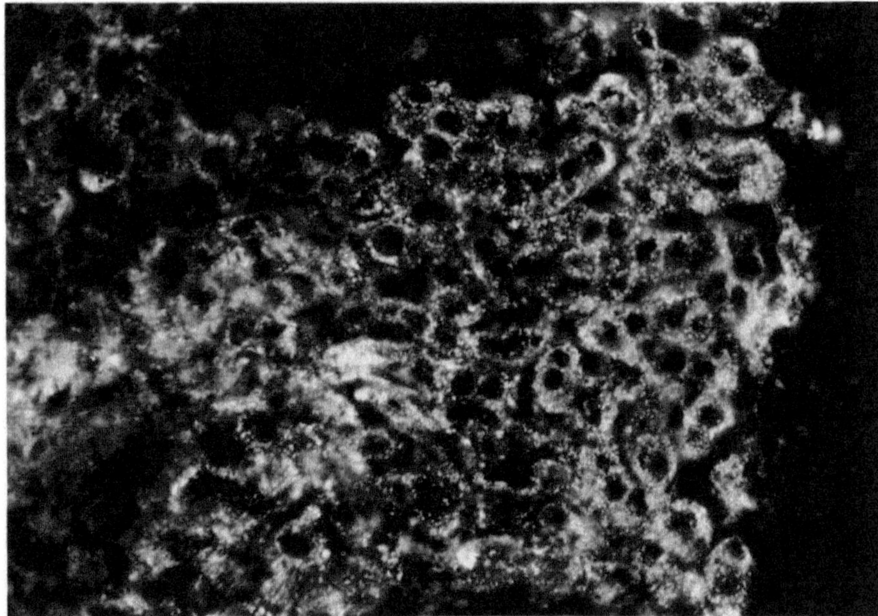

Fig. 6. Tamiami virus infection of *Sigmalon hispidus*. Adrenal gland; focal infection of adrenocortical epithelium without cytopathology. Frozen section, immunofluorescence

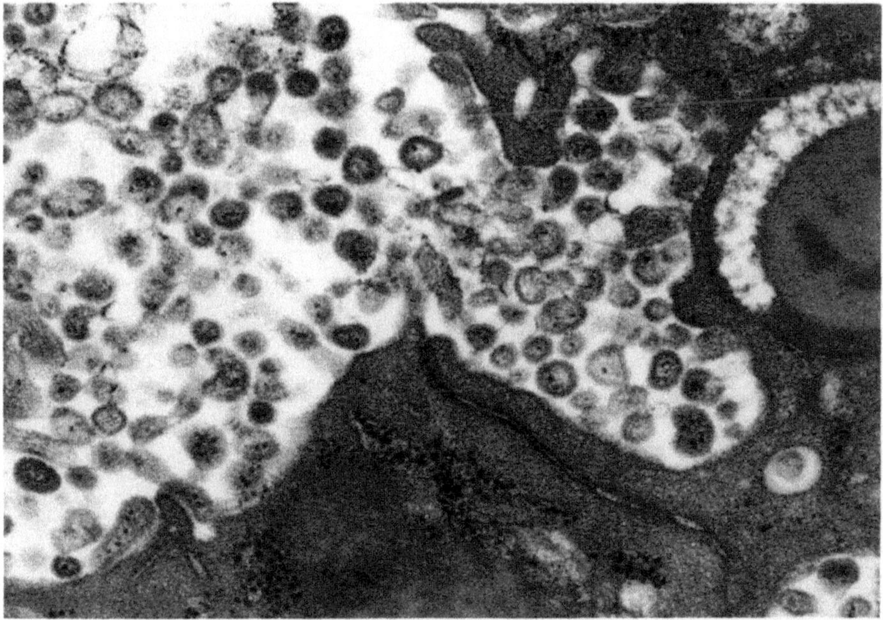

Fig. 7. Machupo virus infection of its natural rodent host, *Calomys callosus*. Salivary gland; virions within a salivary canaliculus between infected mucous epithelial cells. Thin-section electron microscopy

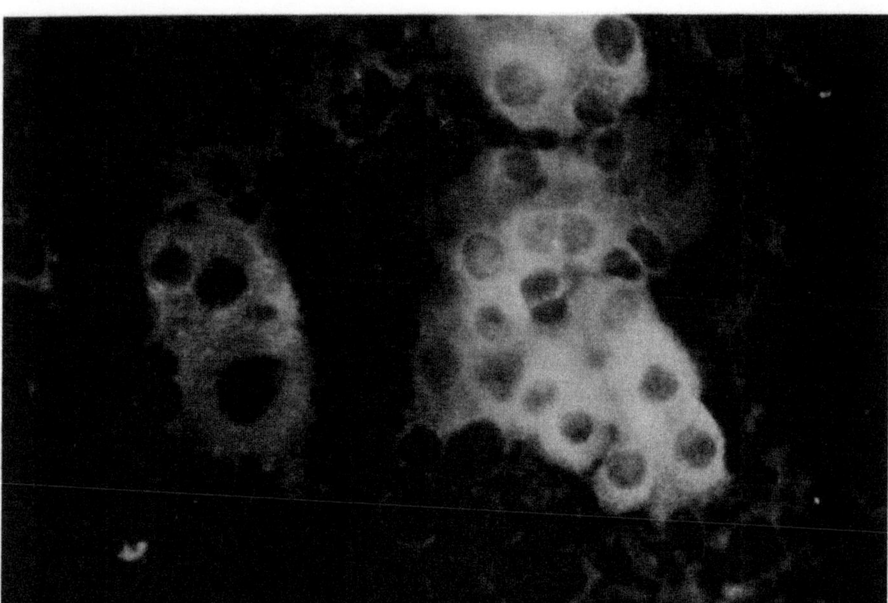

Fig. 8. Lassa virus infection of its natural rodent host, *Mastomys natalensis*. Spleen; infected reticuloendothelial cells in red pulp. Formalin-fixed tissue section, immunofluorescence

Immunosuppressed adult mice develop persistent, noncytopathic infections when inoculated with LCMV. Likewise, immunologically incompetent, genetically athymic "nude" mice inoculated with Junin virus develop persistent viremia and CNS infection without disease (WEISSENBACHER et al. 1983). Mice persistently infected with LCMV have a reduced viral antigenic target for the specific "self-nonself" surveillance activities of the cell-mediated immune system. This reduced target is a result of the decreased expression of LCMV glycoproteins on the surface of infected cells, which in turn contributes to virus survival and persistence (OLDSTONE et al. 1982).

The ability of an infected cell to survive and maintain both the viral infection and its own metabolic and structural requirements during lifelong infection may be related to the ubiquitous phenomenon in arenaviruses of generating defective interfering particles. Conversion of Tacaribe virus preparations that regularly induce a cytopathic effect in vitro to a noncytopathic agent has been accomplished by methods that result in a very high ratio of defective interfering particles to infective virus particles (LOPEZ and FRANZE-FERNANDEZ 1985). The defective interfering particle-rich preparations have a less inhibiting effect on host cell protein and RNA and DNA synthesis; consequently, they do not induce a cytopathic effect, yet infected cells yield a similar amount of infectious progeny as those infected by conventional virus preparations.

There is no evidence that any human arenavirus infection represents an analogue of the silent, persistent, or persistent-tolerant pattern of rodent infections, nor that defective-interfering virus particles play a modulating role in human arenavirus infections. Nevertheless, this kind of virus-host interaction should be sought for more thoroughly in infected humans.

4.2 Noncytocidal Infection with Altered Cell Functions

Noncytocidal arenavirus infections that induce important alterations in the functions of infected cells are difficult to detect. Histopathologic examination of the specific tissues of experimentally infected animals has not offered clues to the presence of physiologic dysfunctions; neither has electron microscopic examination of such tissues. Examples of functional loss in experimental models of arenavirus infection include: (1) decreased production of the neurotransmitter acetylcholine by neuroblastoma cells persistently infected with LCMV, (2) reduced production of growth hormone with consequent dwarfism in mice with persistent LCMV infection of the anterior pituitary gland, and (3) decreased production of insulin with consequent manifestations of diabetes mellitus in mice with persistent LCMV infection of the beta cells of the pancreatic islets of Langerhans (RODRIGUEZ et al. 1983; OLDSTONE et al. 1982, 1984). In each of these models the infected dysfunctioning cells remain intact, but the host animal is progressively affected. This kind of viral effect serves as a reminder, one more in a long series of reminders, that the damage caused by viral infections cannot be equated with cytopathologic effects.

At one point this kind of specific dysfunction caused by persistent arenavirus infection was termed "loss of luxury function", but the nature of these losses makes it clear that the term is inappropriate. Although these losses have not yet been identified in humans infected with arenaviruses, investigations are warranted in all settings where arenavirus diseases are endemic.

4.3 Host-Mediated Immunopathologic Mechanisms

Cytotoxic T-Lymphocyte Mediated Cell Damage. LCMV infection in immunocompetent mice is the classic example of immunopathologic disease – disease caused by host cytotoxic T-lymphocytes as they attack and destroy virus-infected cells that would otherwise remain intact and by-and-large functional (Figs. 9, 10). LCMV immunopathology in certain circumstances is also augmented by circulating immune complexes and by interferon.

LCMV infection in mice is the classic example of histocompatibility-restricted, cell-mediated immune responsiveness. That is, lysis of LCMV-infected cells by cytotoxic T-lymphocytes requires the dual expression on the target cell surface of viral antigen and compatible major histocompatibility antigens (ALLAN and DOHERTY 1985b; ANDERSON et al. 1985; JOHNSON 1985; LEHMANN-GRUBE and LOHLER 1981; OLDSTONE et al. 1985).

This phenomenon has been documented both in vivo and in vitro. Cloned, H-2-restricted, cytotoxic T-lymphocytes specific for LCMV antigen kill infected syngeneic target cells but not uninfected cells or cells infected with vaccinia or Pichinde virus (ANDERSON et al. 1985). For killing to occur, the plasma mem-

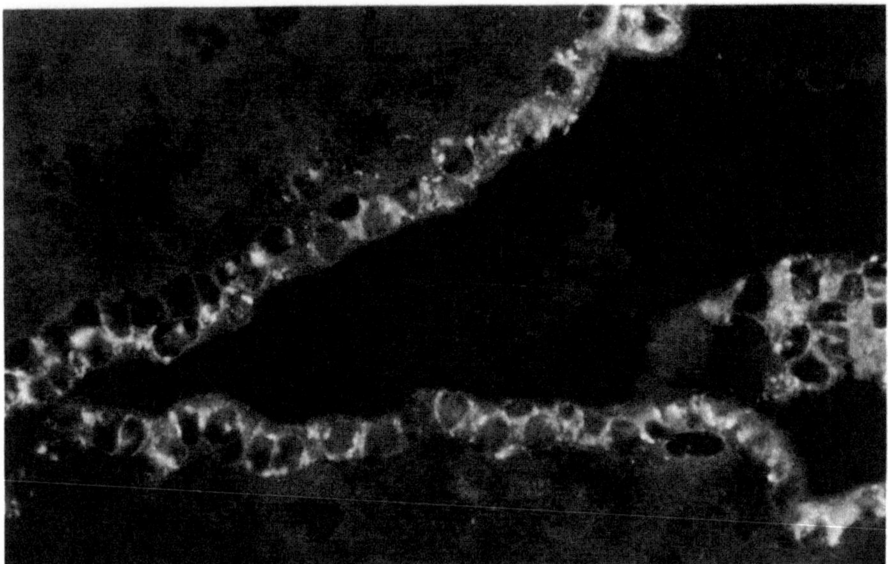

Fig. 9. LCMV infection of *Mus musculus*. Ependyma; infection of epithelium but not underlying brain parenchyma, without cytopathology. Frozen section, immunofluorescence

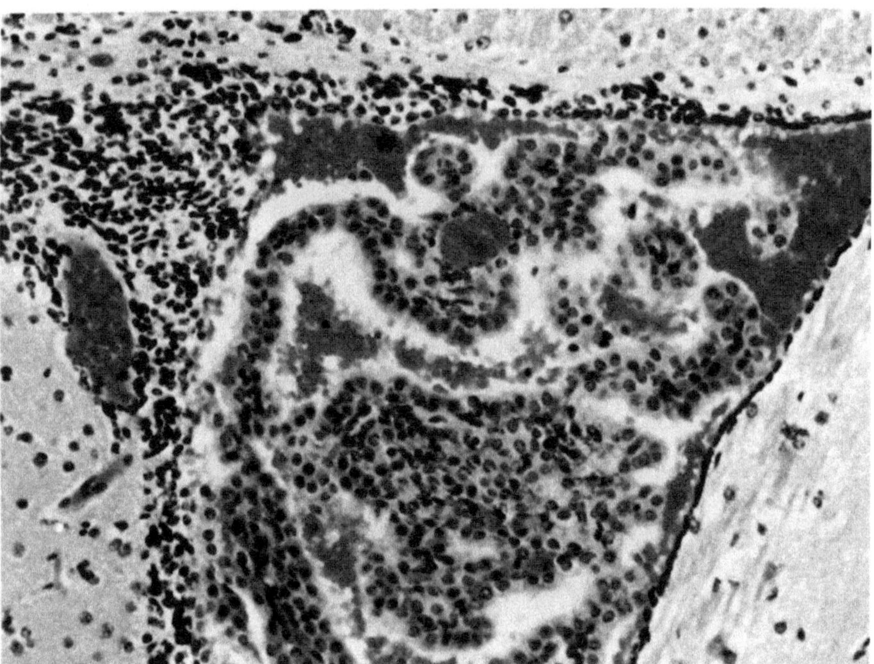

Fig. 10. LCMV infection of *Mus musculus*. Ependyma and choroid plexus; overwhelming immunopathologic damage to infected epithelium by host cytotoxic T-lymphocyte and monocyte/macrophage infiltration. H&E

branes of the cytotoxic T-lymphocytes and the target cells must appear in close apposition; in fact, fingerlike protrusions of the cytotoxic T-lymphocytes invaginate the target cell membrane, and the target cells develop "doughnutlike" lesions in their plasma membranes. The exquisite specificity of this action is indicated by the observation that BALB/c mice mutated so they lack H-2L^d, the only Ir gene in the H-2^d haplotype that influences the cytotoxic T-lymphocyte response to LCMV, exhibit a delay of 24 h in the onset of severe inflammation and clinical signs as compared with normal BALB/c mice (ALLAN and DOHERTY 1985a).

LCMV-infected foci in the CNS of experimentally infected mice, in the choroid plexus, meninges, ependyma, and circumventricular organs, are characterized by lymphohistiocytic infiltrates and ultimately by necrosis of infected cells (WALKER et al. 1975, 1977; CAMENGA et al. 1977). The cellular infiltrates include cells ultrastructurally indistinguishable from cytotoxic T-lymphocytes; these cells have uropods and compound multivesicular bodies (SCHWENDEMANN et al. 1983; MARKER et al. 1984). Shedding of cell fragments, as occurs with cytotoxic T-lymphocyte-target interaction in vitro, has also been observed in vivo.

The immunopathologic nature of LCMV disease in mice was originally proven by using the neuroadapted ARM strain of the virus and the intracerebral route of virus introduction, but the same immunopathologic mechanisms are also active when the viscerotropic WE strain of virus is introduced peripherally. With both virus strains many different immunosuppressive maneuvers prevent the death of adult mice inoculated with virus, but the differences between the events in animals inoculated with ARM versus WE strains are interesting. LCMV ARM, inoculated intracerebrally, establishes critical viral targets for host cytotoxic T-lymphocytes in tissues at the surfaces of the brain – the meninges, choroid plexus, and ependyma. Immunosuppression renders this target expression harmless, but adoptive transfer of sensitized cytotoxic T-lymphocytes restores the pathogenic process in these sites with fatal consequences. LCMV WE, inoculated peripherally, establishes critical viral targets in liver, kidney, and spleen. Adoptive transfer of cytotoxic T-lymphocytes into immunosuppressed infected animals results in hepatitis, splenic necrosis, and death. Although immunosuppression typically results in adult carriers of LCMV, cyclosporin A protects only 50%–70% of mice from death; in the survivors viral clearance and development of neutralizing antibody are not impaired (SARON et al. 1984). Presumably in this case cyclosporin A acts by inhibiting subpopulations of cytotoxic T-lymphocytes, further indicating the functional specificity of the component cells of the cell-mediated immunopathologic response.

Interferon. Interferon increases the susceptibility of LCMV-infected fibroblasts to lysis by H-2-restricted, cytotoxic T-lymphocytes (BUKOWSKI and WELSH 1985). This enhanced lysis correlates with increased cell surface expression of major histocompatibility antigen but not of LCMV antigen. Presumably this observation indicates that the limiting component in the lytic interaction is the histocompatibility antigen, and that physiologic influences upon the expression of antigen might affect the progress of host-mediated disease. Adult mice inoculated intracerebrally with a lethal dose of LCMV and treated with antibody to interferon have been shown to develop high virus titers in organs but benign disease as compared with untreated controls, which develop low virus titers in organs and fatal disease (PFAU et al. 1983). Thus, interferon may play an important role in the pathogenesis of cell damage in murine LCM.

Immune Complex Disease. Murine LCM is the classic example of virus-induced immune complex disease and has served as an important model for human diseases such as hepatitis B carrier-associated glomerulonephritis and arteritis. Persistent infection with circulating nonneutralizing antibody in ratios and quantities that favor binding of complement and deposition in tissues results in the immunopathologic state (OLDSTONE 1984; OLDSTONE et al. 1983). This state is affected by both host and viral genetic factors. Interferon also interacts with immune complex deposits in the genesis of glomerulonephritis. C3H mice, which produce high quantities of interferon, develop very severe ultrastructural lesions when infected with LCMV – their lesions are similar in severity to those induced by exogenous interferon. Strains of mice characterized as low and intermediate interferon responders develop mild and intermediate lesions, respectively. Outbred Swiss mice, characterized as intermediate interferon responders, have higher levels of circulating immune complexes than low interferon-BALB/c mice, and Swiss mice develop much more severe immune complex-associated glomerulonephritis (WOODROW et al. 1982). Antibody to interferon diminishes glomerular damage without affecting levels of circulating immune complexes (PFAU et al. 1983). Thus, both interferon and large quantities of deposited immune complexes appear to play roles in the development of glomerulonephritis.

The role of host-mediated immunopathologic mechanisms, whether by cytotoxic T-lymphocytes, interferon, or immune complexes, in human arenavirus diseases remains unknown. As stated above in Sect. 3.2, there is morphologic evidence that direct viral damage is the basis of the necrotizing hepatitis seen in the disease, but this kind of observation does not preclude the involvement of immunopathologic mechanisms in Lassa fever or other arenavirus hemorrhagic fevers.

4.4 Viral Genetic Composition

Intertypic reassortants between LCMV strains have been used to determine the functional significance of particular virus genes in the pathogenesis of disease. LCMV, like all arenaviruses, has a genome composed of two segments of RNA, a large segment termed L, which probably encodes the virion polymerase, and small segment termed S, which is known to encode the structural proteins of the virus. When intertypic reassortants between a benign virus strain, ARM, and a virulent strain, WE, were inoculated into guinea pigs, only the reassortant with the L segment of WE virus caused fatal acute disease (RIVIERE et al. 1985) and replicated to 10- to 1000-fold higher titers than viruses with the L segment of the benign parent. This observation is not immediately explainable, since the gene products of the S segment, the virion structural proteins, might be expected also to dictate virulence. Nevertheless, the crucial role of the "gene constellation" of the virus in the pathogenesis of acute and persistent arenavirus infections is made clear by these experiments. Understanding of parallel influences in human arenavirus diseases will require novel technical approaches.

4.5 Direct Viral Damage

Direct cell damage caused by arenavirus infection must not be ignored in the overall enthusiasm for cytotoxic T-lymphocyte-mediated pathology, immune complex disease, and noncytopathic infection. In humans it is likely that direct damage to cells is more important than host-mediated immunopathologic and noncytopathic damage. Arenaviruses do cause acute cytopathic effects in some cell cultures, and most may be enumerated by plaque assay. One arenavirus, Tacaribe, has been shown to inhibit host cell macromolecular synthesis of protein, ribosomal RNA, and DNA in association with the induction of cytopathic effect (LOPEZ and FRANZE-FERNANDEZ 1985). The mechanisms by which Tacaribe virus shuts down host cell macromolecular synthesis while viral proteins continue to be synthesized is now known. The molecular basis of arenavirus cytopathic effects in vivo a wait elucidation.

5 Animal Models of Arenavirus Diseases

5.1 Experimental Lassa Virus Infection

Studies of arenaviruses have taught us much about virus-cell interactions and viral immunopathology, but what have they told us about arenavirus diseases themselves? Experimental models of human Lassa fever include squirrel monkeys, rhesus monkeys, and guinea pigs. In these, Lassa virus exhibits an early tropism for lymph nodes, liver, and kidney (WALKER et al. 1975a). In the squirrel monkey at the peak of illness, viral panorganotropism is evident from the high virus titers seen in the lymph nodes, spleen, heart, lungs, liver, pancreas, kidneys, and adrenal glands. Viremia is persistent and lesions include necrosis of the spleen, lymph nodes, and renal tubules, hepatic necrosis and regeneration, myocarditis, focal arteritis, and late in the course, choriomeningoencephalitis (Fig. 11, 12). This model was used to predict the pattern of viral tropism in humans, but it differs in that arteritis and CNS lesions are present; they have not been observed in humans.

The rhesus monkey model has been investigated extensively (JAHRLING et al. 1981; WALKER et al. 1982b; CALLIS et al. 1982; LANGE et al. 1985; FISHER-HOCH et al., submitted for publication). Viremia begins 5 days after inoculation and is higher in fatal cases than in survivors. Virus is recovered from liver, lung, adrenal gland, pancreas, spleen, kidney, and lymph node in titers higher than in blood. Antigens of Lassa virus are detectable in hepatocytes, adrenocortical

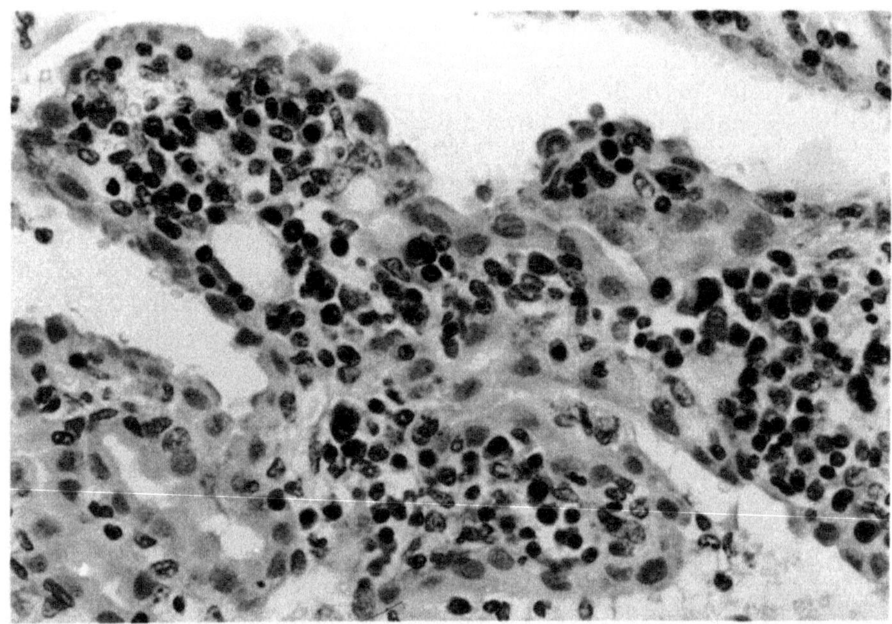

Fig. 11. Lassa virus infection of squirrel monkey (*Saimiri scirrens*). Choroid plexus; mononuclear cell infiltration at 28 days postinfection. H&E

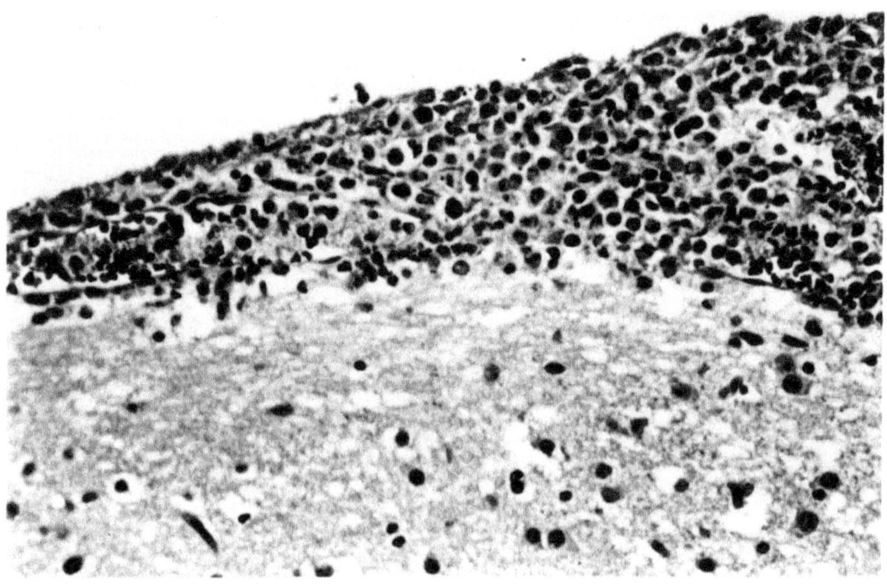

Fig. 12. Lassa virus infection of squirrel monkey. Meninges; mononuclear cell infiltration of leptomeninges at 12 days postinfection. H&E

cells, alveolar septa, lymph nodes, splenic red pulp, renal tubules and glomeruli, thymus, and brain. Endothelium is only minimally infected. The mortality rate is high with pathologic changes including pleural and pericardial effusions, necrosis of hepatocytes and adrenocortical cells, interstitial pneumonia with pulmonary arterial leukocytic aggregates, encephalitis, uveitis, and, as a late complication, arteritis. Serum concentrations of hepatic enzymes are elevated. White blood cell and platelet counts are somewhat depressed. Coagulation parameters including fibrinogen, fibrin degradation products, and platelet and fibrinogen survival times indicate that disseminated intravascular coagulation does not occur in this model. Failure of platelets to aggregate upon exposure to collagen and ADP has been attributed to functional exhaustion analogous to that seen in hemolytic uremic syndrome. Diminished prostacyclin synthesis in aortic tissue collected at necropsy, low 6-keto-prostaglandin F_1, and decreased serum fibronectin suggest damaged or functionally impaired endothelium, although overt endothelial necrosis is not observed histologically.

These and other mechanisms elucidated in monkey models of Lassa fever that might account for the sudden onset of intractable shock in human Lassa fever must be investigated further if treatment of the severely ill patient is to improve. It must be borne in mind, however, that the rhesus and squirrel monkey models differ from humans with Lassa fever in their higher mortality rate, and the prominence of meningoencephalomyelitis, pulmonary vascular lesions, and systemic arteritis.

Outbred guinea pigs infected with Lassa virus develop disseminated infection, pulmonary edema, and foci of necrosis in the liver and heart (WALKER et al. 1975b). Inbred strain 13 guinea pigs experience more severe disease with unformly lethal infection and high viral titers in spleen, lymph node, salivary glands, and lung. These animals also have moderate titers in liver, adrenal gland, kidney, pancreas, and heart (JAHRLING et al. 1982). In these animals Lassa viral antigen is detected by immunofluorescence in lung, spleen, pancreas, kidney, salivary gland, liver, heart, and brain. Interstitial pneumonia occurs in all the animals, but bacterial superinfection has complicated renal and splenic pathologic evaluation. Mild myocarditis is seen in half of these animals, but hepatic, adrenal, and other tissues show disappointingly minimal alterations. This model appears less like human Lassa fever than the monkey models and has yet to yield significant information regarding the pathogenic or pathophysiologic mechanisms of arenaviruses.

5.2 Experimental Junin Virus Infection

In contrast to the modest value of the guinea pig model in investigations of Lassa fever, experimental infection of guinea pigs has produced some interesting ideas regarding the pathogenesis of AHF. Infected animals develop fever, weight loss, thrombocytopenia, and leukopenia, and die on day 11–15 with hemorrhagic lesions (CARBALLAL et al. 1977; KIERSZENBAUM et al. 1970; RIMOLDI and DE BROCCO 1980). Viral infection destroys the bone marrow with release of acid proteases and acid and alkaline phosphatases into the bloodstream, presumably from leukocyte lysosomes. These enzymes may be related to the heat-stable, divalent cation-dependent activator of complement that results in consumption of C4 and depression of total complement activity. This cascade, in turn, possibly leads to alterations in vascular integrity and dysfunction of coagulation mechanisms.

5.3 Experimental Machupo Virus Infection

Experimental Machupo virus infections of rhesus and African green monkeys have been studied as models of human BHF. Rhesus monkeys infected with Machupo virus develop a dose-dependent hemorrhagic disease with fever, rash, lymphadenopathy, splenomegaly, pericardial effusion, depression, anorexia, adipsia, dehydration, hypotension, thrombocytopenia, prolonged partial thromboplastin time, normal or elevated serum fibrinogen levels, and hypoalbuminemia (KASTELLO et al. 1976; SCOTT et al. 1978). There is a high mortality rate. Pathologic lesions include necrosis of hepatocytes, adrenocortical cells, and epithelial cells of the skin, esophagus, mouth, and tongue, necrotizing enteritis, and hemorrhages in skin, heart, brain, and nose. Fibrin thrombi are observed only rarely. Animals that survive the acute phase develop chronic CNS signs and lymphohistiocytic arteritis and meningoencephalomyelitis (MCLEOD et al. 1976; TERRELL et al. 1973). A similar disease is produced in African green monkeys infected with Machupo virus (WAGNER et al. 1977; MCLEOD et al. 1978).

5.4 Experimental Pichinde Virus Infection

The search for an animal model for Lassa fever and South American hemorrhagic fevers, based upon an arenavirus that is minimally pathogenic for humans and upon an animal species that is more available than nonhuman primates, has yielded moderate success. Pichinde virus inoculated peripherally into Golden Syrian hamsters results in an infection that varies in severity according to the genetic background of the host. In inbred HMA strain hamsters of all ages, the infection causes a high mortality rate, but in the outbred LVG strain of hamsters from which the MHA strain was derived, the infection is lethal only in newborns (BUCHMEIER and RAWLS 1977; MURPHY et al. 1977). Adult LVG hamsters, when immunosuppressed with cyclophosphamide and then infected, do develop lethal disease – thus, immunopathologic mechanisms appear less likely than direct viral damage in this model (Figs. 13, 14). Macrophages are the major target cell; there are progressive increases in viral antigen and virus particles (identified ultrastructurally), and necrosis marks the marginal zone and red pulp of the spleen, Kupffer cells, and hepatocytes.

A guinea pig model that relies upon inbred strain 13 animals and a Pichinde virus stock that had been passaged eight times in guinea pigs has been described; the infection is uniformly lethal and the target organs are similar to those in other arenavirus models (JAHRLING et al. 1981). In this model, peak viremia is more than 10^5 TCID$_{50}$/ml of blood. There is lymphopenia, elevated serum

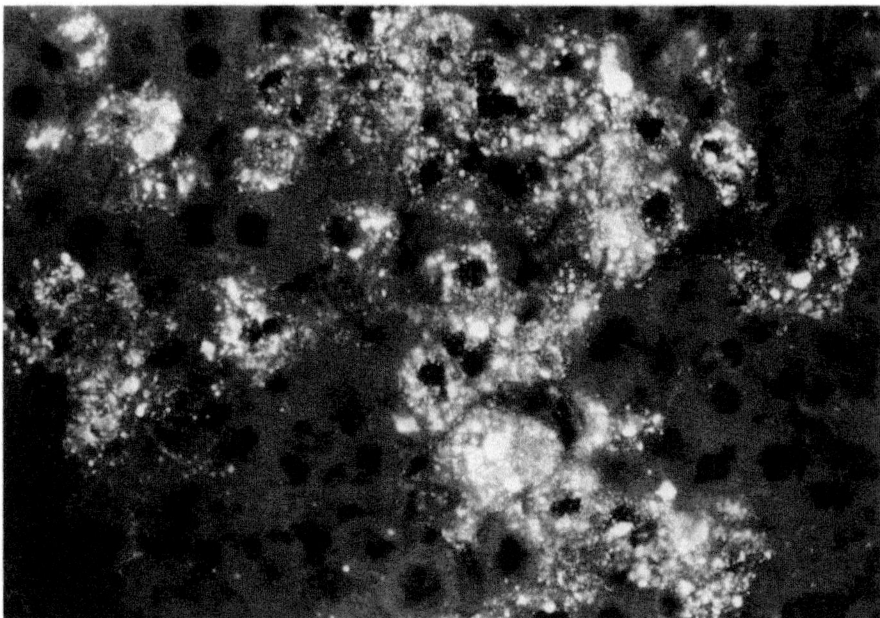

Fig. 13. Pichinde virus infection of MHA hamster. Liver; focal hepatocellular infection at 5 days postinfection, before development of cytopathology. Frozen section, immunofluorescence

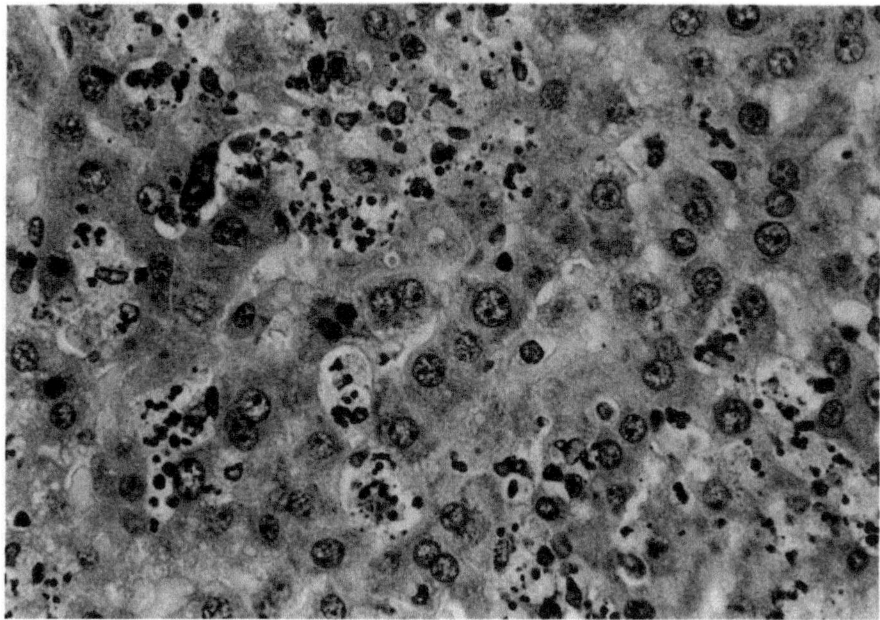

Fig. 14. Pichinde virus infection of MHA hamster. Liver; focal hepatocellular necrosis, without inflammatory response, at 12 days postinfection. H&E

aspartate aminotransferase concentration, hepatocellular, splenic, and adreno-cortical necrosis, and interstitial pneumonia. Viral antigen is detectable by immunofluorescence in hepatocytes, the marginal zone of the periarteriolar lymphatic sheath of the spleen, the adrenal zona fasiculata, the lungs, and the pancreas.

The macrophage as a target cell is an important and recurrent theme in both cytopathic and noncytopathic arenavirus infections (MURPHY et al. 1976; MURPHY et al. 1977; GONZALEZ et al. 1982). The Pichinde virus models reemphasize this tropism, which has also been seen consistently in the benign natural rodent host-virus pairings (MURPHY and WALKER 1978). LCMV infection results in depression of the immune response in both immunologically immature and immunologically intact mice. This immunosuppression is caused by a functional depression of lymphocytes that results from a virus-induced defect in adherent, phagocytic cells (JACOBS and COLE 1976). One of the keys to understanding human arenavirus diseases must lie in the cascade of events that follows infection of the monocyte/macrophage system. It would be satisfying if the mediator or mediators of the terminal shock observed in severe human arenavirus infections were traced to just such a pathogenic cascade. In this regard, the recent studies by FISHER-HOCH, MCCORMICK, and colleagues, reported elsewhere in this volume, point to the central role of endothelial and platelet dysfunction in the shock syndrome seen in Lassa fever. They incriminate a failure in the homeostatic function of prostacyclin upon endothelial integrity and platelet function, in the face of damaging effector substances released by leukocytes and macrophages during the infection. It would be interesting to add the role of leukotrienes to the conceptualization of the damaging cascade of terminal shock; nevertheless, it seems that the answer to the mysterious pathogenic basis for arenavirus hemorrhagic fever and shock is at hand.

References

Allan JE, Doherty PC (1985a) Consequences of a single Ir-gene defect for the pathogenesis of lymphocytic choriomeningitis. Immunogenetics 21:581–589

Allan JE, Doherty PC (1985b) Immune T cells can protect or induce fatal neurological disease in murine lymphocytic choriomeningitis. Cell Immunol 90:401–407

Anderson J, Byrne JA, Schreiber R, Patterson S, Oldstone MBA (1985) Biology of cloned cytotoxic T lymphocytes specific for lymphocytic choriomeningitis virus: clearance of virus and in vitro properties. J Virol 53:552–560

Buchmeier MJ, Rawls WE (1977) Variation between strains of hamsters in the lethality of Pichinde virus infections. Infect Immun 16:413–421

Bukowski JF, Welsh RM (1985) Interferon enhances the susceptibility of virus-infected fibroblasts to cytotoxic T cells. J Exp Med 161:257–262

Callis RT, Jahrling PB, DePaoli A (1982) Pathology of Lassa virus infection in the rhesus monkey. Am J Trop Med Hyg 3:1038–1045

Camenga DL, Walker DH, Murphy FA (1977) Anticonvulsant prolongation of survival in adult murine lymphocytic choriomeningitis. J Neuropathol Exp Neurol 36:9–20

Carballal G, Rodriguez M, Frigerio MJ, Vasquez C (1977) Junin virus infection of guinea pigs: electron microscopic studies of peripheral blood and bone marrow. J Infect Dis 135:367–373

Casals J, Buckley SM (1974) Progress in medical virology. Prog Med Virol 18:111–126

Child PL, MacKenzie RB, Valverde LR, Johnson KM (1967) Bolivian hemorrhagic fever. Arch Pathol 83:434–445

Cossio P, Laguens R, Arana R, Segal A, Maiztegui J (1975) Ultrastructural and immunohistochemical study of the human kidney in Argentine haemorrhagic fever. Virchows Arch [A] 368:1–9

Edington GM, White HA (1972) The pathology of Lassa fever. Trans R Soc Trop Med Hyg 66:381–389

Elsner B, Schwarz E, Mando OG, Maiztegui J, Vilches A (1973) Pathology of 12 fatal cases of Argentine hemorrhagic fever. Am J Trop Med Hyg 22:229–236

Farmer TW, Janeway CA (1942) Infections with the virus of lymphocytic choriomeningitis. Medicine 21:1–63

Fisher-Hoch SP, Mitchell SW, Sasso D, McCormick JB, Lange JV, Ramsey R (submitted for publication) Physiological and immunological disturbances associated with shock in Lassa fever in a primate model. J Infect Dis

Frame JD, Baldwin JM, Gocke DJ, Troup JM (1970) Lassa fever, a new virus disease of man from West Africa. I. Clinical description and pathological findings. Am J Trop Med Hyg 19:670–676

Gonzalez PH, Cossio PM, Arana R, Maiztegui JI, Laguens RP (1980) Lymphatic tissue in Argentine hemorrhagic fever. Arch Pathol Lab Med 104:250–254

Gonzalez PH, Lampuri JS, Coto CE, Laguens RP (1982) In vitro infection of murine macrophages in Junin virus. Infect Immun 35:356–358

Horton J, Hotchin JE, Olson KB, Davies JNP (1971) The effects of MP virus infection in lymphoma. Cancer Res 31:1066–1068

Ishak KG, Walker DH, Coetzer JAW, Gardner JJ, Gorelkin L (1982) Viral hemorrhagic fevers with hepatic involvement: pathologic aspects with clinical correlations. In: Progress in liver disease vol VII. Grune and Stratton, New York

Jacobs RP, Cole GA (1976) Lymphocytic choriomeningitis virus-induced immunosuppression: a virus-induced macrophage defect. J Immunol 117:1004–1009

Jahrling PB, Hesse RA, Eddy GA, Johnson KM, Callis RT, Stephen EL (1980) Lassa virus infection of rhesus monkeys: pathogenesis and treatment with ribavirin. J Infect Dis 141:580–589

Jahrling PB, Hesse RA, Rhoderick JB, Elwell MA, Moe JB (1981) Pathogenesis of a Pichinde virus strain adapted to produce lethal infections in guinea pigs. Infect Immun 32:872–880

Jahrling PB, Smith S, Hesse RA, Rhoderick JB (1982) Pathogenesis of Lassa virus infection in guinea pigs. Infect Immun 37:771–778

Johnson KM (1982) Haemorrhagic fevers. In: Viral disease in southeastern Asia and the western Pacific. Academic Press, Australia

Johnson KM (1985) Arenaviruses. In: Virology. Raven Press, New York

Kastello MD, Eddy GA, Kuehne RW (1976) A rhesus monkey model for the study of Bolivian hemorrhagic fever. J Infect Dis 133:57–62

Kierszenbaum F, Budzko B, Parodi AS (1970) Alterations in the enzymatic activity of plasma of guinea pigs infected with Junin virus. Arch Virusforsch 30:217–223

Knobloch J, McCormick JB, Webb PA, Dietrich M, Schumacher HH, Dennis E (1980) Clinical observations in 42 patients with Lassa fever. Tropenmed Parasitd 31:389–398

Lange JV, Mitchell SW, McCormick JB, Walker DH, Evatt BL, Ramsey RR (1985) Kinetic study of platelets and fibrinogen in Lassa virus-infected monkeys and early pathologic events in mopeia virus-infected monkeys. Am J Trop Med Hyg 34:999–1007

Lehmann-Grube F, Lohler J (1981) Immunopathologic alterations of lymphatic tissues of mice infected with lymphocytic choriomeningitis virus. Lab Invest 44:205–213

Lopez R, Franze-Fernandez M (1985) Effect of Tacaribe virus infection on host cell protein and nucleic acid synthesis. J Gen Virol 66:1753–1761

Maiztegui JI, Laguens RP, Cossio PM, Casanova MB, de la Vega MT, Ritacco V, Segal A, Fernandez NJ, Arana RM (1975) Ultrastructural and immunohistochemical studies in five cases of Argentine hemorrhagic fever. J Infect Dis 132:35–43

Marker O, Nielsen MH, Diemer NH (1984) The permeability of the blood-brain barrier in mice suffering from fatal lymphocytic choriomeningitis virus infection. Acta Neuropathol 63:229–239

McCormick JB, Johnson KM (1978) Lassa fever: historical review and contemporary investigation. In: Pattyn, SR (ed) Ebola virus haemorrhagic fever. Elsevier, New York

McCormick JB, Walker DH, King IJ, Webb PA, Whitfield SG, Johnson KM (1986) Hepatic pathology of Lassa fever. Am J Trop Med Hyg 35:401–407

McLeod CG, Stookey JL, Eddy GA, Scott SK (1976) Pathology of chronic Bolivian hemorrhagic fever in the rhesus monkey. Am J Pathol 84:211–223

McLeod CG, Stookey JL, White JD, Eddy GA, Fry GA (1978) Pathology of Bolivian hemorrhagic fever in the African green monkey. Am J Trop Med Hyg 27:822–826

Molinas FC, Maiztegui JI (1981) Factor VIII:C and factor VIII R:Ag in Argentine hemorrhagic fever. Thromb Haemost (Stuttgart) 46:525–527

Molinas FC, de Bracco MME, Maiztegui JI (1981) Coagulation studies in Argentine hemorrhagic fever. J Infect Dis 143:1–6

Murphy FA, Walker DH (1978) Arenaviruses: persistent infection and viral survival in reservoir hosts. In: Kurstak E, Maramorosch K (eds) Viruses and environment. Academic, New York

Murphy FA, Winn WC, Walker DH, Flemister MR, Whitfield SG (1976) Early lymphoreticular viral tropism and antigen persistence. Lab Invest 34:125–140

Murphy FA, Buchmeier MJ, Rawls WE (1977) The reticuloendothelium as the target in a virus infection. Lab Invest 37:502–515

Oldstone MBA (1984) Virus-induced immune complex formation and disease: definition, regulation, importance. In: Notkins AL, Oldstone MBA (eds) Concepts in Viral Pathogenesis. Springer, New York Berlin Heidelberg Tokyo

Oldstone MBA, Buchmeier MJ (1982) Restricted expression of viral glycoprotein in cells of persistently infected mice. Nature 300:360–362

Oldstone MBA, Sinha YN, Blount P, Tishon A, Rodriguez M, von Wedel R, Lampert PW (1982) Virus-induced alterations in homeostasis: alterations in differentions in differentiated functions of infected cells in vivo. Science 218:1125–1127

Oldstone MBA, Tishon A, Buchmeier MJ (1983) Virus-induced immune complex disease: genetic control of C1q binding complexes in the circulations of mice persistently infected with lymphocytic choriomeningitis virus. J Immunol 130:912–918

Oldstone MBA, Southern P, Rodriguez M, Lampert P (1984) Virus persists in cells of islets of Langerhans and is associated with chemical manifestations of diabetes. Science 224:1440–1443

Oldstone MBA, Ahmed R, Byrne J, Buchmeier MJ, Riviere Y, Southern P (1985) Virus and immune resposes: lymphocytic choriomeningitis virus as a prototype model of viral pathogenesis. Br Med Bull 41:70–74

Pfau CJ, Gresser I, Hunt KD (1983) Lethal role of interferon in lymphocytic choriomeningitis virus-induced encephalitis. J Gen Virol 64:1827–1830

Rimoldi MT, de Bracco MM (1980) In vitro inactivation of complement by a serum factor present in Junin virus-infected guinea pigs. Immunology 39:159–164

Riviere Y, Ahmed R, Southern PJ, Buchmeier MJ, Oldstone MBA (1985) Genetic mapping of lymphocytic choriomeningitis virus pathogenicity: virulence in guinea pigs in associated with the L RNA segment. J Virol 55:704–709

Rodriguez M, von Wedel RJ, Garrett RS, Lampert PW, Oldstone MBA (1983) Pituitary dwarfism in mice persistently infected with lymphocytic choriomeningitis virus. Lab Invest 49:48–53

Saron M-F, Shidani B, Guillon J-C, Truffa-Bachi P (1984) Beneficial effect of cyclosporin A on the lymphocytic choriomeningitis virus infection in mice. Eur J Immunol 14:1064–1066

Sarrat H, Camain R, Baum J, Robin Y (1972) Diagnostic histopathologique des hepatites dues an virus Lassa. Bull Soc Pathol Exot Filiales 65:642–650

Schwendemann G, Lohler J, Lehmann-Grube F (1983) Evidence for cytotoxic T-lymphocyte-target cell interaction in brains of mice infected intracerebrally with lymphocytic choriomeningitis virus. Acta Neuropathol (Berl) 61:183–195

Scott SK, Hickman RL, Lang CM, Eddy GA, Hilmas DE, Spertzel RO (1978) Studies of the coagulation system and blood pressure during experimental Bolivian hemorrhagic fever in rhesus monkeys. Am J Trop Med Hyg 27:1232–1239

Smadel LE, Green RH, Paltauf RM, Gonzales TA (1942) Lymphocytic choriomeningitis: two human fatalities following an unusual febrile illness. Proc Soc Exp Biol Med 49:683–686

Terrell TG, Stookey JL, Eddy GA, Kastello MD (1973) Pathology of Bolivian hemorrhagic fever in the rhesus monkey. Am J Pathol 73:477–494

Wagner FS, Eddy GA, Brand OM (1977) The African green monkey as an alternate primate host for studying Machupo virus infection. Am J Trop Med Hyg 26:159–162

Walker DH, Wulff H, Murphy FA (1975a) Experimental Lassa virus infection in the squirrel monkey. Am J Pathol 80:261–268

Walker DH, Wulff H, Lange JV, Murphy FA (1975b) Comparative pathology of Lassa virus infection in monkeys, guinea-pigs, and *Mastomys natalensis*. Bull WHO 52:523–534

Walker DH, Murphy FA, Whitfield SG, Bauer SP (1975c) Lymphocytic choriomeningitis: ultrastructural pathology. Exp Mol Pathol 23:245–265

Walker DH, Camenga DL, Whitfield S, Murphy FA (1977) Anticonvulsant prolongation of survival in adult murine lymphocytic choriomeningitis. J Neuropathol Exp Neurol 36:21–40

Walker DH, McCormick JB, Johnson KM, Webb PA, Komba-Kono G, Elliott LH, Gardner JJ (1982a) Pathologic and virologic study of fatal Lassa fever in man. Am J Pathol 107:349–356

Walker DH, Johnson KM, Lange JV, Gardner JJ, Kiley MP, McCormick JB (1982b) Experimental infection of rhesus monkeys with Lassa virus and a closely related arenavirus, Mozambique virus. J Infect Dis 146:360–368

Warkel RL, Rinaldi CF, Bancroft WH, Cardiff RD, Holmes GE, Wilsnack RE (1973) Fatal acute meningoencephalitis due to lymphocytic choriomeningitis virus. Neurology 23:198–203

Webb PA, Justines G, Johnson KM (1975) Infection of wild and laboratory animals with Machupo and Latino viruses. Bull WHO 52:493–499

Weissenbacher MC, Calello MA, Quintans CJ, Panisse H, Woyskowsky NM, Zannoli VH (1983) Junin virus infection in genetically athymic mice. Intervirology 19:1–5

Winn WC, Walker DH (1975) The pathology of human Lassa fever. Bull WHO 52:535–545

Winn WC, Monath TP, Murphy FA, Whitfield SG (1975) Lassa virus hepatitis. Arch Pathol 99:599–604

Woodrow D, Ronco P, Riviere Y, Moss J, Gresser I, Guillon JC, Morel-Maroger L, Sloper JC, Verroust P (1982) Severity of glomerulonephritis induced in different strains of suckling mice by infection with lymphocytic choriomeningitis virus: correlation with amounts of endogenous interferon and circulating immune complexes. J Pathol 138:325–336

Subject Index

MIX
Papier aus verantwortungsvollen Quellen
Paper from responsible sources
FSC® C105338

FSC
www.fsc.org

If you have any concerns about our products,
you can contact us on
ProductSafety@springernature.com

In case Publisher is established outside the EU,
the EU authorized representative is:
Springer Nature Customer Service Center GmbH
Europaplatz 3, 69115 Heidelberg, Germany

Printed by Libri Plureos GmbH
in Hamburg, Germany